Comparative Life Cycle Assessment of Industrial Multi-Product Processes

Von der Fakultät für Maschinenwesen der
Rheinisch-Westfälischen Technischen Hochschule Aachen
zur Erlangung des akademischen Grades eines Doktors
der Ingenieurwissenschaften genehmigte Dissertation

vorgelegt von

Johannes Jung

Berichter: Univ.-Prof. Dr.-Ing. André Bardow
Prof. Reinout Heijungs, PhD

Tag der mündlichen Prüfung: 23. August 2013

Aachener Beiträge zur Technischen Thermodynamik Band 2
Johannes Jung
Comperative Life Cycle Assesement of Industrial Multi-Product Processes

ISBN: 978-3-86130-471-5

Bibliografische Information der Deutschen Bibliothek
Die Deutsche Bibliothek verzeichnet diese Publikation in der Deutschen Nationalbibliografie; detaillierte bibliografische Daten sind im Internet über http://dnb.ddb.de abrufbar.

Vertrieb:

1. Auflage 2014
© Wissenschaftsverlag Mainz GmbH - Aachen
Süsterfeldstr. 83, 52072 Aachen
Tel. 0241/87 34 34
Fax 0241/87 55 77
www.Verlag-Mainz.de

ISSN: 2198-4832

Herstellung:

Druck und Verlagshaus Mainz GmbH Aachen
Süsterfeldstraße 83
52072 Aachen
Tel. 0241/87 34 34
Fax 0241/87 55 77
www.DruckereiMainz.de

Satz: nach Druckvorlage des Autors
Umschlaggestaltung: Druckerei Mainz

printed in Germany
D82 (Diss. RWTH Aachen University, 2013)

'Nur Einzigartigkeit entzieht sich der Vergleichbarkeit.'

Dr. F.P. Rinnhofer

'Only uniqueness escapes comparability.'

Dr. F.P. Rinnhofer

Preface

Die vorliegende Arbeit entstand im Rahmen meiner Tätigkeit als wissenschaftlicher Mitarbeiter am Lehrstuhl für Technische Thermodynamik der RWTH Aachen. Meinem Doktorvater, Prof. Dr.-Ing. André Bardow, danke ich sehr herzlich für die Förderung dieser Arbeit. Die häufigen Diskussionen und vielfältigen Anregungen haben maßgeblich zum Gelingen der Arbeit beigetragen. Herrn Prof. Dr. Reinout Heijungs danke ich für die Übernahme des Koreferats und die motivierenden Gespräche im Vorfeld. Herrn Prof. Dr.-Ing. Matthias Wessling gilt mein Dank für die Übernahme des Vorsitzes der Prüfungskommission.

Meinem Kollegen Niklas von der Assen danke ich besonders für die intensiven Diskussionen und konstruktiven Hinweise, mit denen er meine Arbeit kontinuierlich begleitet hat. Weiterhin danke ich Patrick Eichstädt für die Implementierungen in Matlab, Arne Kätelhön für das Korrekturlesen sowie Sarah Postels für die Unterstützung bei der schriftlichen Ausarbeitung.

Ich hatte das Glück, in meiner Arbeit ein aktuelles Fallbeispiel untersuchen zu dürfen. Meinen Projektpartnern von Bayer MaterialScience, der Uhde GmbH, und den Siegfried Jacob Metallwerken danke ich daher für die gute Kooperation.

Meinem langjährigen Bürokollegen und Freund Philip Voll danke ich sehr für die gemeinsame Zeit in unserem „Office“, die von harter Arbeit, sportlichen Herausforderungen, gegenseitiger Motivation und freundschaftlichen Ratschlägen geprägt war. Weiterhin danke ich meinen langjährigen Wegbegleitern Christoph Kausch, Stefan Kirschbaum, Matthias Lampe, Benjamin Meyer und André Sternberg für die freundschaftliche Arbeitsatmosphäre.

Meinen Eltern Marlies und Norbert Jung danke ich sehr für ihr Vertrauen und die Unterstützung auf meinem bisherigen Lebensweg. Abschließend danke ich besonders herzlich meiner Frau Evelien für ihre Bereitschaft, für mich nach Aachen zu ziehen und mich unentwegt zu unterstützen.

Aachen, im Januar 2014 *Johannes Jung*

Contents

List of Figures

List of Tables

List of matrices and vectors

List of symbols

a	electrolysis current efficiency 8
α	number of avoided burden processes that are aggregated to a single avoided burden process 51
β	number of functional flows in a multi-functional unit process 54
F	FARADAY constant 8
f_i	specific FARADAY constant substance i 8
γ_i	normalized sensitivity coefficient of the ith elementary flow 91
K	total number of allocation criteria with corresponding allocation factors used to calculate average allocation factors 102
Λ	total number of alternative processes in comparative LCA 66
M_i	molar mass of substance i 8
η_k	normalized sensitivity coefficient of the kth impact flow 91
Ω	total number of aggregation factors used to calculate an average aggregation factor 96
φ	variable for the functional flow of a multi-functional process that is initially not included in the functional unit 53
σ	standard deviation 37
σ^2	variance 37
σ_q	normalized sensitivity coefficient of the qth scaling factor 91
U_{anode}	electrochemical potential of the anode reaction 7
U_{cathode}	electrochemical potential of the cathode reaction 7

U_{cell}	operating cell voltage 7
$U_{\text{cell,ODC}}$	operating cell voltage of ODC process 11
U_{min}	decomposition voltage, minimum cell voltage 7
$U_{\text{min,ODC}}$	decomposition voltage of ODC process, minimum voltage cell voltage of ODC process 11
U_{ODC}	electrochemical potential of the ODC reaction 11
w_{el}	specific electricity demand 8
$w_{\text{el,ODC}}$	specific electricity demand of an ODC process 11
z_i	number of electrons transfered per ion of substance i 8
ζ	relative uncertainty contribution 94

Acronyms

AP	acidification potential 129
av	average 6
CA	chlor-alkali 116
CFP	carbon footprint 35
CHP	combined heat and power generation 24
el	*electricity* 68
EOL	end-of-life 14
EP	eutrophication potential 129
eq	equivalent 30
EU-Commission	European Commission 15
EU-ETS	European emissions trading scheme 1
exp	expanded 28
FU	functional unit 17
GB	Great Britain 135
GER	Germany 135
GHG	greenhouse gas 9
GW	global warming 119
GWI	global warming impact 21
GWP	global warming potential 21
he	*heat* 69
HTP	human toxicity potential 129
ini	initial 48

ISO	International Organization for Standardization 15
LCA	life cycle assessment 1
LCI	life cycle inventory 15
LCIA	life cycle impact assessment 15
MC	Monte-Carlo 41
mf	multi-functional 25
mp	main product 73
NG	natural gas 112
NL	the Netherlands 135
NO	Norway 135
ODC	oxygen depolarized cathode 9
ODP	ozone depletion potential 129
PC	polycarbonate 5
POCP	photochemical ozone creation potential 129
PP	polypropylene 114
PTFE	polytetrafluoroethylene 129
PUR	polyurethane 5
PVC	polyvinyl chloride 5
RU	Russia 135
SETAC	Society of Environmental Toxicology and Chemistry 15
STC	standard cathode 110
UNEP	United Nations Environment Programme 15

Kurzfassung

Die Anforderungen an die Umweltverträglichkeit industrieller Prozesse sind in den vergangenen Jahrzehnten deutlich gestiegen. Daher sollen Umweltauswirkungen neuer Prozesse möglichst frühzeitig bewertet werden. Die Analyse von Umweltauswirkungen über den gesamten Lebenszyklus eines Prozesses kann mit einer Ökobilanz (engl. Life Cycle Assessment, LCA) erfolgen. Innerhalb von Ökobilanzen treten häufig Allokationsprobleme auf. Ein Allokationsproblem entsteht z.B., wenn die Umweltauswirkungen eines Prozesses auf mehrere Produkte des Prozesses aufgeteilt werden sollen. Eine solche Aufteilung ist an vielen Stellen notwendig, z.B. beim Vergleich alternativer Prozesse mit mehreren Produkten, bei denen die Produkte nur teilweise übereinstimmen. Zur Lösung eines Allokationsproblems stehen drei alternative Methoden zur Wahl: Systemerweiterung, Substitution und Allokation. Nach der Auswahl einer dieser drei Methoden sind zusätzliche Entscheidungen innerhalb der jeweiligen Methode erforderlich. Alle Wahlmöglichkeiten erhöhen die Unsicherheiten und verringern somit die Akzeptanz einer Ökobilanz.

Ein internationaler Standard und mehrere Praxisleitfäden liefern Empfehlungen für die Auswahl geeigneter Methoden zur Lösung von Allokationsproblemen. Bisher sind diese Empfehlungen allerdings nicht für vergleichende Ökobilanzen alternativer Prozesse konkretisiert. Außerdem werden Unsicherheiten aufgrund der Wahlmöglichkeiten bei der Lösung von Allokationsproblemen selten in systematischen Analysen berücksichtigt: In einigen Beispielen werden Szenarien mit alternativen Lösungsmethoden einander gegenübergestellt. Teilweise werden die Unsicherheiten innerhalb dieser Szenarien noch mit rechenintensiven stochastischen Verfahren untersucht. Allerdings sind diese Ansätze weder standardisiert noch in die Rechenalgorithmen von Software-Werkzeugen zur Ökobilanzierung eingebunden.

In der vorliegenden Arbeit wird eine konkrete Anleitung zur Auswahl der Lösungsmethode bei Allokationsproblemen in vergleichenden Ökobilanzen vorgestellt. Darüber hinaus wird eine analytische Methode zur systematischen Erfassung und Analyse der Unsicherheiten aufgrund von Allokationsproblemen entwickelt.

Die Anleitung zur Auswahl der Lösungsmethode bei Allokationsproblemen in vergleichenden Ökobilanzen setzt voraus, dass die Produkte der zu vergleichenden Prozesse auf Basis wirtschaftlicher Überlegungen in Haupt- und Nebenprodukte eingeteilt werden. Unterscheiden sich zwei zu vergleichende Prozesse in ihren Hauptprodukten, sollte die Systemerweiterung angewendet werden. Liegt der Unterschied in den Nebenprodukten, sollte die Anwendung einer standortspezifischen Substitution der Allokation vorgezogen werden.

Für ein analytisches Verfahren zur Analyse der Unsicherheiten aufgrund von Allokationsproblemen wird zunächst der bestehenden Formelsatz zur Berechnung von Ökobilanzen erweitert. Die Erweiterung des Formelsatzes umfasst kontinuierliche Parameter, welche die diskreten Wahlmöglichkeiten bei der Lösung von Allokationsproblemen modellieren. Auf Basis des erweiterten Formelsatzes werden Sensitivitätskoeffizienten berechnet, welche die Sensitivität einer Ökobilanz hinsichtlich der Auswahlmöglichkeiten bei der Lösung von Allokationsproblemen beschreiben. Die Anwendung einer Fehlerfortpflanzung erster Ordnung auf den erweiterten Formelsatz ermöglicht es letztendlich, die Auswirkungen der Unsicherheiten aufgrund von Allokationsproblemen systematisch und rechnergestützt zu untersuchen.

Die Anwendbarkeit der entwickelten Methoden wird in einer vergleichenden Ökobilanz zweier alternativer Prozesse aus der chemischen Industrie gezeigt: Ein neues energiesparendes Verfahren zur Chlor-Alkali-Elektrolyse produziert Chlor und Natronlauge. Die zu vergleichende, etablierte Technologie produziert allerdings neben Chlor und Natronlauge zusätzlich noch Wasserstoff. Daher entsteht bei der vergleichenden Ökobilanz ein Allokationsproblem. Die Methode zur Lösung des Allokationsproblems kann mit der entwickelten Anleitung ausgewählt werden. An Standorten, wo Wasserstoff chemisch genutzt wird, erfolgt eine Systemerweiterung. An Standorten, wo der Wasserstoff thermisch verwertet wird, erfolgt eine Substitution. In beiden Fällen bleiben die gesamten Umweltauswirkungen des neuen Verfahrens geringer als die Auswirkungen durch die etablierte Technologie. Eine Unsicherheitsanalyse mit Hilfe der entwickelten Methodik zeigt, dass hier die Unsicherheiten aufgrund des Allokationsproblems gegenüber Datenunsicherheiten nicht zu vernachlässigen sind. Die in dieser Arbeit entwickelten Methoden können zukünftig in bestehende Software-Werkzeuge eingebunden werden und erhöhen die Akzeptanz von Ökobilanzen.

Chapter 1

Introduction

Smoking chimneys are a symbol for environmental impacts of industrial processes. Indeed, industrial processes are major contributors to environmental problems such as global warming (Bernstein et al., 2007). Beyond emission-related problems, industrial processes deplete limited resources because they require raw materials. Raw materials are directly linked to costs, emission-related impacts cause indirect expenditures, e.g., through the European emissions trading scheme (EU-ETS) for greenhouse gas emissions. Therefore, industrial enterprises seek to reduce costs by reducing environmental impacts of their processes.

Two well-known strategies for reducing environmental impacts of industrial processes are process integration and recycling (Friedler, 2010; Klemes et al., 2010; Ayres, 1997). Process integration establishes interconnections between formerly separate processes by utilizing co-products (El-Halwagi, 2006; Foo, 2012). Process integration thereby relies on unit processes with multiple products, so-called multi-product processes. Similarly, recycling uses waste as raw material for new products (De La Mantia and Mantia, 2002; Lund, 2001). But neither process integration nor recycling guarantee reduced environmental impacts. E.g., recycling may cause more impacts than waste disposal (Hischier et al., 2005; Cattlin and Wang, 2012). Decision makers thus need a holistic method for comparing environmental impacts from multi-product processes.

This work investigates methods for comparisons of industrial multi-product processes. The environmental impacts of multi-product processes can be analyzed using life cycle assessment (LCA). LCA studies all environmental impacts of all processes involved in a product's entire life cycle. Due to its holistic approach, LCA identifies shifting of environmental problems between processes and between different types of environmental impacts. Results of LCA-studies can thus help avoiding such problem shifting.

But multi-product processes cause a methodological problem in LCA: products of a multi-product process share the environmental impacts of that process. If only one product from a multi-product process is studied, parts of the process' environmental impacts have to be related to the studied product. In LCA, this situation is referred to as the multi-functionality problem. Comparisons of multi-product processes frequently create such multi-functionality problems. If a power plant that generates only electricity is compared to combined heat and power generation, it remains questionable whether the basis of the comparison should be electricity generation or electricity and heat generation. Depending on the basis of comparison, the environmental impacts of combined heat and power generation have to be partitioned among the products heat and electricity.

LCA provides three methods to fix such multi-functionality problems. One method allocates the environmental impacts of a multi-product process to its single products. For combined heat and power generation, the impacts could, e.g., be allocated based on the energy content of the products electricity and heat. A second method subtracts environmental impacts avoided by co-production of the product not studied. Those subtracted impacts refer to a process which is avoided because the product from the multi-functional process is utilized. For combined heat and power, environmental impacts from a coal-fired boiler for heat generation could be subtracted. At last, a third method chooses the basis of comparison so that all products from a multi-functional process are included and thereby fixes a multi-functionality problem. This work aims to provide a procedure how to fix multi-functionality problems for comparisons of multi-product processes.

Unfortunately, the methods for fixing multi-functionality problems are subject to potentially ambiguous choices. E.g., allocation of environmental impacts among electricity and heat could also be based on exergy content instead of energy content. LCA-practitioners thus have to choose an allocation criterion. Likewise, one could also subtract environmental impacts from a gas-fired instead of a coal-fired boiler for heat generation. Here, the practitioner has to select among alternative processes. The subjective choices thus introduce uncertainty to LCA. Those uncertainties often irritate decision makers seeking precise solutions for reducing environmental impacts. Moreover, a systematic method for analyzing the uncertainties due to fixing multi-functionality problems is lacking.

This work aims to rigorously quantify uncertainties due to fixing multi-functionality problems using analytical error propagation. The resulting framework allows to compare the often important influence of fixing multi-functionality problems to other uncertainties in a systematic manner. Decision makers could ultimately base their judgement on a single holistic and computationally efficient uncertainty analysis that could be incorporated in existing software tools.

1.1 Structure of this thesis

This thesis begins with a motivating example for comparative LCA of multi-product processes from the chemical industry: a recently developed technology for chlor-alkali electrolysis. The novel technology promises large reductions in electricity demand but has also fewer usable co-products than existing technologies. Comparing environmental impacts of the novel technology with those of existing ones thus creates a multi-functionality problem. This example is described in chapter 2. The open question from chapter 2 is whether the novel technology has lower environmental impacts than existing technologies.

In chapter 3, the holistic LCA approach for environmental impact assessment is introduced. In section 3.1 the general concept, nomenclature, and computational structure of LCA is presented. The three methods for fixing multi-functionality problems are described in section 3.2. Existing tools for uncertainty analysis in LCA are discussed in section 3.3.

In chapter 4, an existing matrix formulation for LCA calculations is expanded to include methods for fixing multi-functionality problems. The derived equations are a fundamental premise for applying analytical error propagation to analyze uncertainties due to fixing multi-functionality problems.

In chapter 5, a systematic procedure for fixing multi-functionality problems in comparative LCA of multi-product processes is provided. The procedure is based on existing recommendations for fixing multi-functionality problems in non-comparative LCA. In chapter 6, analytical error propagation is introduced for analysis of uncertainties due to fixing multi-functionality problems in LCA.

The methods derived in chapter 4 to 6 are applied in the comparative LCA-study of chlor-alkali electrolysis in chapter 7. This chapter serves two purposes: The previously developed methods are applied in a real-world example and the application-oriented question raised in chapter 2 is answered. The thesis is completed with conclusion for LCA-methodology and the chlor-alkali industry in chapter 8.

Readers primarily interested in the LCA-study of chlor-alkali electrolysis technologies may consecutively read chapter 2 and chapter 7. Readers interested in the methodological problems and solutions developed in this work may focus on chapter 3 to chapter 6.

Chapter 2

Motivating example: chlor-alkali electrolysis

2.1 Chlor-alkali electrolysis - an energy-intensive multi-product process

The chemical industry relies heavily on chlorine and caustic soda: in Europe, e.g., more than 55 % of the chemical industry's turnover depend on the two essential chemical commodities chlorine and caustic soda (EuroChlor, 2011). Nearly one third of the produced chlorine is used to manufacture polyvinyl chloride (PVC) (Fauvarque, 1996). Further important products are polycarbonates (PCs), polyurethanes (PURs), and titanium dioxide (TiO_2). Caustic soda is used in aluminum production, oil refineries, and during manufacturing of paper, textiles, and soaps (Fauvarque, 1996). A detailed overview of organic, inorganic, and pharmaceutical products requiring chlorine or caustic soda is given in O'Brien et al. (2005a) and the chlorine tree (American Chemistry Council, 2013).

Chlorine and caustic soda are mentioned in the same context because they are co-produced in the so-called chlor-alkali electrolysis. Generally, electrolysis is the non-spontaneous electrochemical decomposition of a substance driven by direct electric current (Hamann et al. (2007) on p. 5). Chlor-alkali electrolysis refers to the decomposition of sodium chloride (NaCl) and water (H_2O) to chlorine (Cl_2), sodium hydroxide (NaOH), and hydrogen (H_2) (Schmittinger et al. (2006) on p. 8). The reaction equation of chlor-alkali electrolysis is

$$2\,\mathrm{NaCl} + 2\,\mathrm{H_2O} \longrightarrow \mathrm{Cl_2} + 2\,\mathrm{NaOH} + \mathrm{H_2}. \tag{2.1}$$

Cl_2 and H_2 are produced as gases. In contrast, NaCl and NaOH are both in aqueous solution: NaCl in water is called brine, the aqueous solution of NaOH is caustic soda. Chlorine, caustic soda, and hydrogen are always produced jointly due to the fixed stoichiometry of the chlor-alkali electrolysis reaction (Equation 2.1). Therefore, the chlor-alkali electrolysis is intrinsically a multi-product process.

Chlor-alkali electrolysis is responsible for more than 95 % of the worldwide chlorine and caustic soda production (EuroChlor, 2012). The worldwide production was 68 Mt chlorine and 76 Mt caustic soda in 2008 (Jörissen et al., 2011). The electricity demand for worldwide chlor-alkali electrolysis is enormous: EuroChlor (2012) estimated an average (av) electricity demand of $w_{\mathrm{el,av}} = 3400\,\mathrm{kW\,h/t\ Cl_2}$. Using those figures, the worldwide annual electricity consumption of chlor-alkali electrolysis is estimated to

$$3400\,\frac{\mathrm{kW\,h}}{\mathrm{t\ Cl_2}} \cdot 68\,\mathrm{Mt\ Cl_2} = 231.2\,\mathrm{TWh}. \tag{2.2}$$

The electricity consumption is comparable to the annual electricity demand of countries such as Mexico (225.76 TWh), South Africa (240.09 TWh), or Australia (226.96 TWh, International Energy Agency (2012)). Chlor-alkali electrolysis is thus known as one of the most energy-intensive industrial processes (Neelis et al., 2007).

Understanding the origin of the high electricity demand requires knowledge about the technical design of chlor-alkali electrolysis plants and the underlying electrochemistry. In the following paragraphs, the basic design of the state-of-the-art electrolysis plant is briefly described. Thereafter, the fundamental electrochemical relationships are presented to explain the electricity demand of chlor-alkali electrolysis.

The state-of-the-art technology for chlor-alkali electrolysis is the membrane process (European Commission, 2001). The membrane process takes place in a plant unit called electrolyzer. The electrolyzer consists of identical, serially connected electrolysis cells. The schematic design and operation of a single cell is shown in Figure 2.1. A so-called membrane cell is divided into an anode compartment and a cathode compartment by an ion-exchange membrane. The membrane separates the gaseous products Cl_2 and H_2 preventing an explosive reaction of these two gases.

The anode compartment is fed with NaCl and H_2O (brine). The Cl^--ions in the brine are discharged at the positively charged anode forming gaseous chlorine according to

$$2\,\mathrm{Cl^-} \longrightarrow \mathrm{Cl_2} + 2\,\mathrm{e^-}. \tag{2.3}$$

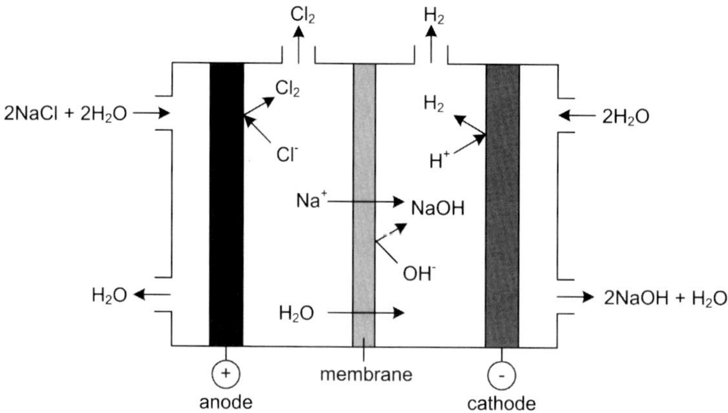

Figure 2.1: Schematic design and operation of membrane electrolysis cell

The electric field forces hydrated Na^+-ions to move through the ion-exchange membrane into the cathode compartment. The cathode compartment is fed with H_2O. At the cathode, H_2 and OH^--ions are generated from H_2O:

$$2\,H_2O + 2\,e^- \longrightarrow H_2 + 2\,OH^-. \tag{2.4}$$

The selective membrane suppresses the migration of OH^--ions into the anode compartment. Instead, the OH^--ions react with the migrated Na^+-ions to sodium hydroxide according to

$$Na^+ + OH^- \longrightarrow NaOH. \tag{2.5}$$

NaOH is solved in water forming caustic soda (Figure 2.1). More detailed information on the design of membrane cells can be found in (O'Brien et al., 2005b).

The reactions given in Equations 2.3 and 2.4 only occur when a voltage is applied to anode and cathode of each cell. This operating cell voltage U_{cell} is the cause for the electrolysis' electricity demand.

The laws of thermodynamics determine a minimum cell voltage U_{min}. This minimum cell voltage is given by the electrochemical potentials of both electrode reactions (cf. Equations 2.3 and 2.4). The difference between both potentials U_{anode} and U_{cathode} is the minimum cell voltage:

$$U_{\text{min}} = U_{\text{anode}} - U_{\text{cathode}}. \tag{2.6}$$

The electrochemical potentials can be calculated from the physical properties of the participating substances and the reaction conditions. Typical reaction conditions for a membrane process (cf. O'Brien et al. (2005c) on pp. 196-198) yield

$$U_{\text{min}} = 1.2322\,\text{V} - (-0.9944\,\text{V}) = 2.2266\,\text{V}. \tag{2.7}$$

The operating cell voltage U_{cell} is higher than U_{min} because of overpotentials and ohmic losses. Overpotentials exist at the surfaces of the electrodes and the membrane. Ohmic losses occur in brine, caustic soda, the membrane, and the electrodes (cf. Schmittinger et al. (2006) on p. 45). The operating voltage of a modern membrane cell is thus $U_{\text{cell}} \approx 3.0\,\text{V}$ (cf. O'Brien et al. (2005b) on p. 415 and O'Brien et al. (2005c) on p. 206).

The electricity demand can be calculated from the cell voltage U_{cell} using FARADAY's law (cf. O'Brien et al. (2005c) on p. 163). The specific electricity demand w_{el} for any electrolysis product is

$$w_{\text{el}} = \frac{U_{\text{cell}}}{a \cdot f_i}. \tag{2.8}$$

Here, f_i is the specific FARADAY constant for substance i and a is the current efficiency. A specific FARADAY constant f_i is calculated from $f_i = \frac{M_i}{F \cdot z_i}$, where F is the FARADAY constant, M_i the molar mass of substance i, and z_i the number electrons transferred per ion of substance i. For chlorine, the specific Faraday constant becomes

$$f_{\text{Cl}_2} = \frac{M_{\text{Cl}_2}}{F \cdot z_{\text{Cl}_2}} = \frac{70.906\,\text{g/mol}}{96\,485\,\text{A s/mol} \cdot 2} = 1.3228\,\frac{\text{kg}}{\text{kA h}}. \tag{2.9}$$

The current efficiency a accounts for losses which prevent all chlorine from reacting at the anode (cf. O'Brien et al. (2005c) on p. 167). For a typical current efficiency of $a = 97\,\%$ (e.g., O'Brien et al. (2005a) on p. 436) and a cell voltage of $U_{\text{cell}} = 3.0\,\text{V}$ (cf. O'Brien et al. (2005c) on pp. 186 and 206), the specific electricity demand of a membrane process becomes

$$w_{\text{el}} = \frac{U_{\text{cell}}}{a \cdot f_{\text{Cl}_2}} = \frac{3.0\,\text{V}}{97\,\% \cdot 1.3228\,\text{kg/(kA h)}} = 2338\,\frac{\text{kW h}}{\text{t Cl}_2}. \tag{2.10}$$

Even though the electricity demand is specified per mass unit chlorine, it describes the required electricity for joint production of chlorine, caustic soda, and hydrogen according to Equation 2.1.[1]

The electricity demand in Equation 2.10 is considerably lower than the worldwide average of chlor-alkali electrolysis ($w_{\mathrm{el,av}} = 3400\,\mathrm{kW\,h/t\,Cl_2}$, cf. EuroChlor (2011)). The reason for this discrepancy is that only half of the worldwide electrolysis plants employ the described membrane process (EuroChlor, 2012). The remaining capacities use less efficient mercury or diaphragm processes. Both mercury and diaphragm processes are outdated and not installed in new plants because of environmental restrictions and inefficient operation behavior (EuroChlor, 2011). For this reason, mercury and diaphragm processes are not further elaborated in this work. The interested reader is referred to O'Brien et al. (2005c), O'Brien et al. (2005d), O'Brien et al. (2005b), and Schmittinger et al. (2006) for detailed discussions of the mercury and diaphragm processes.

The production costs of chlor-alkali electrolysis depend up to 60 % on the electricity costs (Schmittinger et al. (2006) on p. 87). Rising electricity prices push the chlor-alkali industry to develop techniques that lower the electricity demand.

An environmental motivation for lowering the electricity demand are greenhouse gas (GHG) emissions from electricity generation. The electricity required for chlor-alkali electrolysis is often generated in fossil-fueled power plants. These power plants emit large amounts of GHG emissions which are partly induced by chlor-alkali electrolysis. Reducing the electricity demand of chlor-alkali electrolysis thus reduces GHG emissions, from electricity generation.

A promising technology for reducing the specific electricity demand of chlor-alkali electrolysis is introduced in the following section 2.2.

2.2 Reducing electrolysis' electricity demand with oxygen depolarized cathodes (ODCs)

An approach for reducing electricity demand of chlor-alkali electrolysis is using an oxygen depolarized cathode (ODC) (Yeager and Bindra, 1980; Moussallem et al., 2008).

[1] In some references, the specific electricity demand of chlor-alkali electrolysis is actually given per mass unit caustic soda, cf. O'Brien et al. (2005b). That specific electricity demand can be converted to an equivalent specific demand per mass unit chlorine using only the molar masses of the components.

Since 2011, a pilot electrolyzer equipped with ODCs is operated in Germany (Bayer, 2011). The idea of ODCs is employing a cathode reaction with a lower electrochemical potential than Equation 2.4. A lower electrochemical potential causes a lower minimum cell voltage following Equation 2.6 and ultimately reduces the electricity demand calculated from Equation 2.8.

The reaction at an ODC prevents the formation of hydrogen by reducing oxygen (O_2) according to

$$O_2 + 2\,H_2O + 4\,e^- \longrightarrow 4\,OH^-. \tag{2.11}$$

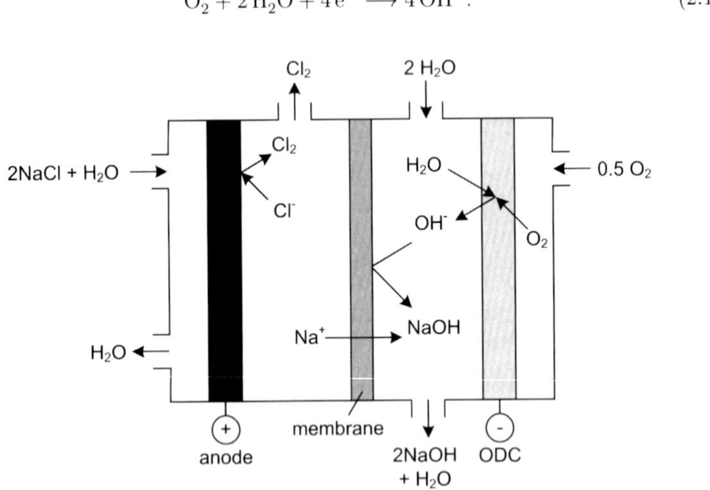

Figure 2.2: Schematic design and operation of electrolysis cell using an ODC

The schematic design and operation of an electrolysis cell using an ODC is illustrated in Figure 2.2. In the cathode compartment, the ODC separates the gaseous oxygen and the liquid caustic soda. The ODC is designed to allow oxygen diffusion into a reaction zone while withstanding the hydrostatic pressure of the liquid caustic soda (Sugiyama et al., 2003; Pinnow et al., 2011). The anode compartment and the ion-exchange membrane remain unchanged compared to the membrane process. The anode reaction is thus also equal to Equation 2.3. Consequently, the overall reaction of a chlor-alkali electrolysis with an ODC becomes

$$2\,NaCl + H_2O + 0.5\ O_2 \longrightarrow Cl_2 + 2\,NaOH. \tag{2.12}$$

A Chlor-alkali electrolysis using ODCs is called ODC-process in this work.

The ODC reaction in Equation 2.11 has a standard potential U_{ODC} at 25 °C of $U_{\mathrm{ODC}} = 0.4\,\mathrm{V}$ (cf. O'Brien et al. (2005e) on p. 1466). Under the operating conditions assumed in O'Brien et al. (2005e), the minimum voltage $U_{\mathrm{min,ODC}}$ of an ODC process becomes

$$U_{\mathrm{min,ODC}} = 1.02\,\mathrm{V} \tag{2.13}$$

Overpotentials and ohmic losses increase the operating cell voltage to $U_{\mathrm{cell,ODC}} \approx 2.0\,\mathrm{V}$ (O'Brien et al., 2005e; Moussallem et al., 2008; Bulan et al., 2009; Jörissen et al., 2011). The corresponding electricity demand $w_{\mathrm{el,ODC}}$ becomes

$$w_{\mathrm{el,ODC}} = \frac{U_{\mathrm{cell,ODC}}}{a \cdot f_{\mathrm{Cl_2}}} = \frac{2.0\,\mathrm{V}}{97\,\% \cdot 1.3228\,\mathrm{kg/(kA\,h)}} = 1558\,\frac{\mathrm{kW\,h}}{\mathrm{t\ Cl_2}}. \tag{2.14}$$

Here, the current efficiency $a = 97\,\%$ is assumed to remain the same compared to the membrane process (Bayer Material Science, 2010a).[2] Compared to a membrane process, the ODC-process reduces the electricity demand by 33 % (cf. Equation 2.10). To obtain this advantage, the ODC-process requires pure oxygen as input and does not produce hydrogen anymore.

As a consequence of these differences, environmental impact reductions such as lower GHG emissions are not directly deducible from the advantage in electricity demand: membrane and ODC-processes require different resources during operation (oxygen) and manufacturing (noble metals) and do not both produce hydrogen. In the following section 2.3, the research problem of comparing environmental impacts from membrane and ODC-processes is formulated.

2.3 Problem statement: environmental impacts of membrane and ODC-processes

At first glance, the ODC-process seems environmentally favorable because of its reduced electricity demand. But manufacturing ODCs involves noble metals and is more complicated than standard cathodes of the membrane process (Moussallem et al.,

[2]The assumption for the current efficiency should be updated as soon as validated data from ODC plants is available.

2008). Moreover, the ODC-process requires pure oxygen and does not produce hydrogen. But hydrogen from the membrane process is typically utilized: more than 90 % of hydrogen produced by the European chlor-alkali industry is used as fuel or chemical commodity (EuroChlor, 2012). A sound comparison of the environmental impacts has to integrate all these differences between the membrane and ODC-process.

So far, a full assessment of the potential environmental impacts of the ODC-process is missing. Jörissen et al. (2011) correctly state that roughly 30 % of CO_2 -emissions related to electricity generation can be saved by the ODC-process. These authors include oxygen production in their study but do not further discuss the effect of omitted hydrogen production and ODC manufacturing. Moussallem et al. (2009) compare the energy demand of membrane and ODC-processes. They include utilization of hydrogen from the membrane process for energy recovery through combustion or fuel cells. Their results show that the ODC-process still has a lower electricity demand if it is compared to the membrane process including energy recovery through hydrogen utilization. Still, these authors do not discuss hydrogen utilization as chemical commodity and focused on energetic and economic performance indicators rather than on environmental impacts.

A method including both energetic indicators and environmental impacts is LCA. LCA has not been applied to the ODC-process so far. The first goal of this work is thus a LCA-based comparison of environmental impacts from the membrane and ODC-processes. Beyond this application-specific goal, two methodological problems arise from applying LCA to compare the membrane and ODC-processes:

First, including the utilization of hydrogen in the LCA-study is expected to cause a multi-functionality problem that needs to be fixed. Methods for fixing multi-functionality problems in LCA exist, but guidelines for their application to compare multi-product processes are lacking. Second, fixing multi-functionality problems introduces uncertainty to the LCA-results. So far, systematic analytical methods for analyzing uncertainties due to fixing multi-functionality problems are missing.

These methodological problems are related to the utilization of the co-product hydrogen, in other words to the multi-functionality of both electrolysis processes. To understand the methodological problems in detail, multi-functionality in LCA is introduced in the following chapter 3. A brief description of the general concept of LCA (section 3.1) is necessary to discuss the background of the two methodological problems in detail: fixing multi-functionality problems for comparative LCA of multi-product processes (section 3.2) and analysis of uncertainties due to fixing multi-functionality problems (section 3.3).

Chapter 3

Multi-functionality and uncertainty analysis in life cycle assessment

Motivated by the comparison of chlor-alkali electrolysis in chapter 2, this chapter provides a comprehensive overview of existing methods dealing with multi-functionality and uncertainties due to multi-functionality in LCA. In section 3.1, the fundamental basics of LCA are briefly introduced. These basics are necessary to understand the application of LCA for analyzing environmental impacts of industrial processes. The section includes the general concept of LCA (section 3.1.1), the terminology used in this work (section 3.1.2), and the computational structure of LCA (section 3.1.3).

In section 3.2, multi-functionality in LCA is introduced. In this section, different types of multi-functional processes are described in section 3.2.1 before a formal definition of the multi-functionality problem is given in section 3.2.2. The existing methods for fixing multi-functionality problems are reviewed in section 3.2.3. Based on this review, the first methodological problem of this work aims at fixing multi-functionality problems for comparative LCA of multi-product processes. This problem is derived in section 3.2.4.

All methods for fixing the multi-functionality problem introduce uncertainties. The uncertainties and methods for analyzing in LCA are described in section 3.3. First, the uncertainties due to fixing multi-functionality problems are formulated and compared with other types of uncertainty in LCA (section 3.3.1). Existing methods for uncertainty analysis are reviewed focusing on the application to uncertainties due to multi-functionality (section 3.3.2). Based on this overview, the second methodological problem of this work calls for an analytical approach for analyzing uncertainties due to fixing multi-functionality problems in LCA. This problem is formulated in section 3.3.3 and concludes this chapter.

3.1 Life cycle assessment (LCA)

3.1.1 General concept of LCA

The brief overview of *life cycle assessment* presented in this section follows detailed descriptions in existing textbooks (e.g., Guinée et al. (2002), Baumann and Tillman (2004), or Klöpffer and Grahl (2009)).

LCA investigates the environmental impacts of processes throughout an entire product life cycle. This holistic approach is illustrated in Figure 3.1: a product life begins with extraction of raw materials from the environment. These raw materials are manufactured into a product. The product is packed, transported and then used for a specific purpose. After being used, a product is either recycled or faces its end-of-life (EOL), i.e., it is disposed as waste in the environment.

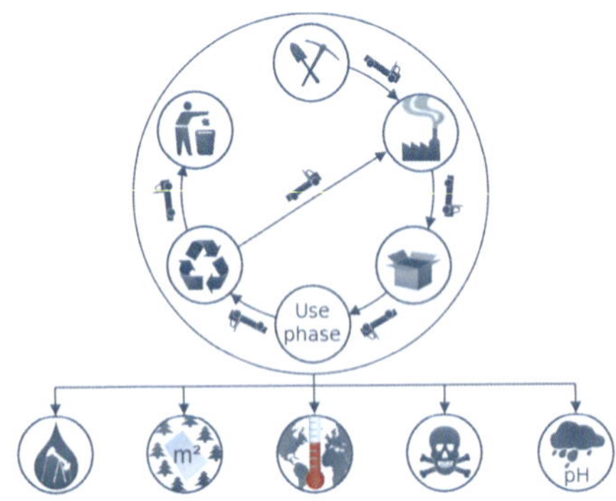

Figure 3.1: Schematic illustration of a product life cycle and potential environmental impacts (von der Assen, 2011)

A product life cycle usually has direct impacts on the environment, because the extracted raw materials typically differ from the disposed wastes. More environmental impacts occur during a product life cycle through further extraction of materials from the environment (e.g., for packaging of the product) and emissions to the environment (e.g., from electricity generation for operation of manufacturing processes). LCA aims to include all interactions of a product life cycle with the environment. These interactions are aggregated to environmental impacts. Examples for environmental impacts are fossil resource depletion, land use, toxicity, ozone depletion, or global warming (Figure 3.1).

The holistic approach of LCA is capable of analyzing changes in a product life cycle. Such changes often aim at improving product characteristics, or at reducing manufacturing costs and environmental impacts. But changes in one life cycle stage (e.g., manufacturing) may shift problems to other life cycle stages (e.g., recycling). In the same way, reductions of one environmental impact (e.g., global warming) may cause an increase in another impact (e.g., ozone depletion). Identifying both kinds of problem shifting is LCA's characteristic feature.

The concept of LCA was born in the early 1970s, when environmental problems such as air pollution or resource scarcity aroused public attention (Boustead, 1972; Hunt and Welch, 1974; Oberbacher et al., 1974). In the following "decades of conception" (Guinee et al., 2011), several international organizations and institutions developed a methodological framework for LCA (Hunt et al., 1996; Oberbacher et al., 1996; Boustead, 1996; Fink, 1997). Those regionally parallel developments caused diverging procedures and terminologies. The methodological divergence was overcome during the 1990s, when the Society of Environmental Toxicology and Chemistry (SETAC) and the International Organization for Standardization (ISO) coordinated the international harmonization of LCA. The harmonization culminated in the international standards (ISO, 2006a,b) for LCA.

The standard's methodological framework divides LCA into four procedural steps as shown in Figure 3.2: *goal and scope definition*, *inventory analysis*, *impact assessment*, and *interpretation*. First, the intended application and underlying assumptions of a LCA-study are stated as goal and scope. In the life cycle inventory (LCI)-analysis, data is collected and the inputs and outputs of a product life cycle are calculated. These inputs and outputs are associated with environmental impacts in life cycle impact assessment (LCIA). Finally, interpretation and discussion of these environmental impacts in the context of the goal and scope concludes a LCA-study (ISO, 2006a).

Even though ISO (2006a,b) specify *what* should be done in these four steps, many questions remain open *how* to do it (Reap et al., 2008b,a). Therefore, international organizations such as the United Nations Environment Programme (UNEP) and the European Commission (EU-Commission) have published detailed guiding documents which address many questions on how to perform LCA (Udo de Haes and van Rooijen, 2005; Sonnemann and Vigon, 2011; European Commission, 2010b). These guiding documents are useful supplements of ISO (2006a) and available free of charge.

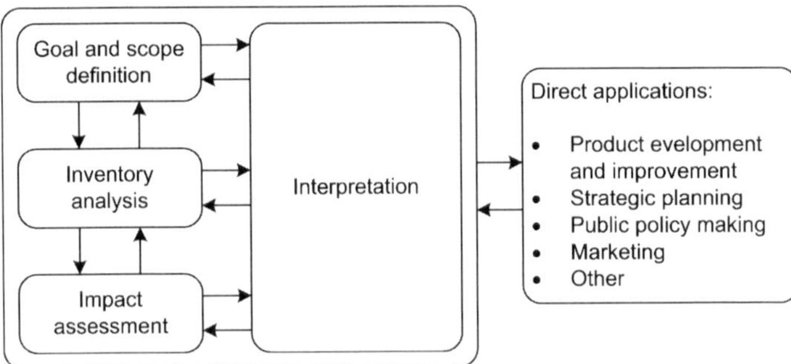

Figure 3.2: Procedural steps for LCA following ISO (2006a): goal and scope definition; life cycle inventory analysis; life cycle impact assessment; interpretation.

Moreover, LCA has developed into an active research field. Thus, e.g., methodological questions are frequently discussed from a scientific point of view (Rebitzer et al., 2004; Pennington et al., 2004; Reap et al., 2008b,a; Finnveden et al., 2009). However, the number and variety of available literature also caused the divergence in terminologies to survive. For this reason, the following section 3.1.2 describes the terminology used in this work.

3.1.2 Terminology and definitions for LCA

The terminology of this work aligns with the definitions of the standard (ISO (2006a) on pp. 7-14). Where needed, own definitions are added to avoid any confusion with other literature.

A product life cycle can be modeled as a *process system* interacting with the system environment. Figure 3.3 shows an exemplary process system. The dashed lines represent the system boundary separating the process system and the system environment. A process system is composed of *unit processes*. Unit processes describe transformations of energy and materials within the product life cycle. The transformation within each unit process is quantified by its input and output data. These input and output data are material or energy flows.

Material and energy flows can be exchanged between unit processes and also between unit processes and the environment. Flows between a unit process and the environment are named *elementary flows*. Flows exchanged between unit processes

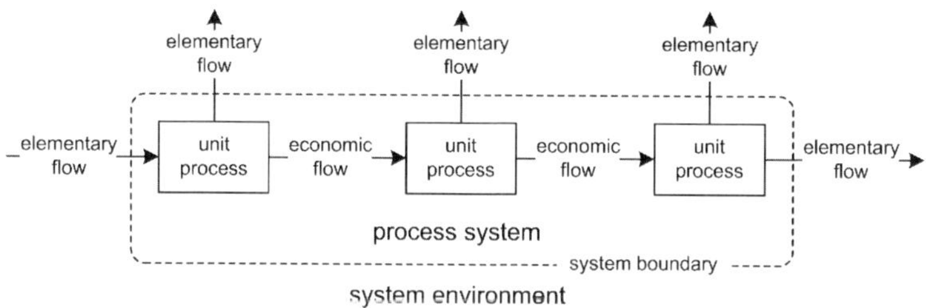

Figure 3.3: Process flow diagram of a process system in LCA following ISO (2006a)

are called *economic flows*. Economic flows are further characterized by their economic value (Guinée et al., 2009): economic flows with a positive economic value are called *products*. In contrast, economic flows with a negative economic value are called *wastes*.

The subject of a LCA-study is defined in the *functional unit* (ISO (2006a), section 4.1.4 on p. 14). The functional unit (FU) describes the "quantified performance of a process system for use as a reference unit" (ISO (2006a), section 3.20 on p. 10).[1] This performance is quantified by *functional flows*. Functional flows are economic flows that correspond to the function of a unit process. The functional flows of industrial processes are either product output flows or waste input flows. A functional unit such as "producing a specified amount of a product" corresponds with a product output flow. A functional unit such as "processing a specified amount of a waste" relates to a waste input flow.

The functional unit is particularly important for *comparative LCA*. Comparative LCA aims at comparing environmental impacts from *alternative products* or from *alternative processes* producing the same product. For comparative LCA of both alternative products and processes, the comparison shall use the same functional unit (ISO, 2006a).

Functional unit and system boundaries are defined during the goal and scope step of LCA (ISO (2006a), section 5.2.1.2 on p. 23). The economic and elementary flows of all unit process within a processes system are collected during the LCI-analysis. In the subsequent LCIA, the elementary flows are aggregated to environmental impacts.

[1] ISO (2006a) uses the additional term *reference flow* for the "measure of outputs from processes in a given system required to satisfy the functional unit". In this work, the functional unit and reference flow are equivalent because the considered functional units are directly defined as measurable outputs of processes.

LCA concludes with the interpretation of these environmental impacts. The interpretation commonly reflects the results in the context of the goal and scope, states limitations and provides recommendations for decision makers.

Both LCI-analysis and LCIA deal with large amounts of data: a product life cycle is easily composed of several hundred unit processes. Each unit process may again include hundreds of input and output flows. Handling these data requires efficient computational algorithms. Therefore, the computational structure of LCA is briefly introduced in the following section 3.1.3.

3.1.3 Computational structure of LCA

Two alternative computational concepts were developed over the past decades: the sequential[2] and the matrix-based approach (Schmidt and Schorb, 1996; Heijungs and Suh, 2002).

The sequential approach uses a central unit process (Ciroth et al., 2004). The central unit process' functional flows usually correspond to the functional unit. Flow balances of all economic flows are sequentially calculated for all unit processes up- and downstream from the central unit process. These economic flow balances yield scaling factors for each unit process. These scaling factors are used for calculating the corresponding elementary flows of each unit process. Finally, all elementary flows are summed up and aggregated to environmental impacts.

The sequential approach is implemented in the LCA software tools GaBi and Umberto used by many LCA-practitioners (PE International, 2013; ifu, 2013). Still, the sequential approach suffers from computational problems once process systems include loops of economic flows or unit processes with multiple product flows (Heijungs and Suh, 2002; Suh and Huppes, 2005). Both loops of economic flows and unit processes with multiple products occur frequently for industrial process systems. The matrix-based approach overcomes these disadvantages (Heijungs and Suh, 2002). Therefore, the sequential approach is not further used in this work and the matrix-based LCA is introduced in more detail in the following section 3.1.3.1.

[2]The sequential approach is also called network method (Ciroth et al., 2004) or process flow diagram (Suh and Huppes, 2005).

3.1.3.1 Matrix-based LCA

The matrix-based approach models a process system as a set of matrices and vectors (Heijungs, 1994, 1997). These matrices and vectors form a linear equation system if all unit processes can be scaled linearly. Solving the linear equation system using matrix inversion yields the environmental impacts of a process system. A very detailed description of the matrix-based approach was given by Heijungs and Suh (2002). The matrix-based approach is also implemented in software tools, e.g., Simapro (Goedkoop et al. (2006) on p. 15) and CMLCA (Heijungs, 2013a). This section introduces the basic equations and terminology of matrix-based LCA. Readers who are familiar with matrix-based LCA may directly skip to section 3.2.

A virtual process system is used for illustration. The process flow diagram of the process system is shown in Figure 3.4). The process system consists of two unit processes: mining (process 1) and electricity generation (process 2). In this illustration, the input and output flows of both processes are scaled to produce one default unit of product output. The mining process extracts 1.2 kg carbon from the environment to deliver 1 kg coal while emitting 0.73 kg CO_2 to the environment. The electricity generation process generates 1 kWh electricity from 0.31 kg coal and emits 1.14 kg CO_2 to the environment.[3]

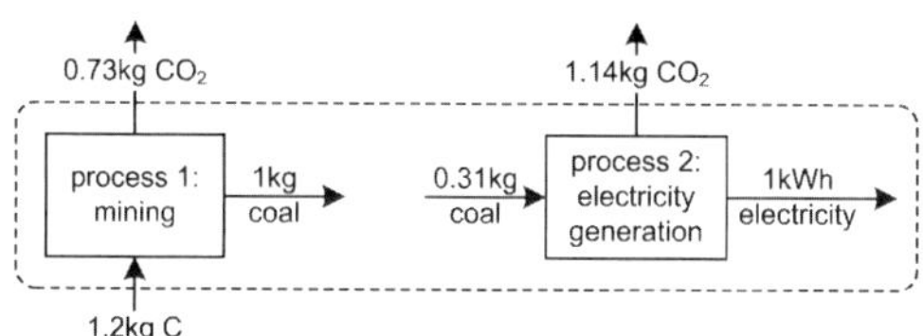

Figure 3.4: Process flow diagram for a process system with two unit processes: mining and electricity generation.

In matrix-based LCA, each unit process is modeled by a *process vector* $\mathbf{p}_j$. A process vector comprises all input and output flows of a unit process. In line with Heijungs and Suh (2002), input flows are negative and output flows are positive in this work. For the example shown in Figure 3.4, the two process vectors $\mathbf{p}_1$ for mining and $\mathbf{p}_2$ for electricity generation are

[3]The figures of the illustrating examples are virtual numbers throughout this work.

$$\mathbf{p}_1 = \begin{pmatrix} 1.00\,\text{kg coal} \\ 0.00\,\text{kWh electricity} \\ -1.20\,\text{kg C} \\ 0.73\,\text{kg CO}_2 \end{pmatrix} \text{ and } \mathbf{p}_2 = \begin{pmatrix} -0.31\,\text{kg coal} \\ 1.00\,\text{kWh electricity} \\ 0.00\,\text{kg C} \\ 1.14\,\text{kg CO}_2 \end{pmatrix}.^4 \tag{3.1}$$

All process vectors of a system are arranged to a *process matrix* **P**:

$$\mathbf{P} = \left(\mathbf{p}_1 \mid \mathbf{p}_2 \right). \tag{3.2}$$

The process matrix **P** represents a product's LCI. The process matrix **P** can be partitioned into two submatrices **A** and **B** by separating economic and elementary flows in the process vectors:

$$\mathbf{P} = \left(\frac{\mathbf{A}}{\mathbf{B}} \right). \tag{3.3}$$

The submatrix **A** comprises all economic flows and is called *technology matrix*. In contrast, submatrix **B** contains all elementary flows and is named *intervention matrix*. For the illustrating example, technology matrix **A** and intervention matrix **B** are

$$\mathbf{A} = \begin{pmatrix} 1 & -0.31 \\ 0 & 1 \end{pmatrix} \text{ and } \mathbf{B} = \begin{pmatrix} -1.20 & 0 \\ 0.73 & 1.14 \end{pmatrix}. \tag{3.4}$$

Here, each process has one functional flow: mining produces coal, electricity generation electricity.

In matrix-based LCA, the functional unit is given as *final demand vector* **f**. The final demand vector has the same row dimension as the technology matrix **A**. It is a sparse vector because only functional flows that correspond with the functional unit have non-zero values. In the illustration example, the functional unit is "producing 1000 kWh electricity from coal". The corresponding final demand vector **f** includes the functional flow of the electricity generation-process, i.e.,

$$\mathbf{f} = \begin{pmatrix} 0 \\ 1000 \end{pmatrix}. \tag{3.5}$$

[4] To improve readability, units and flow names are dropped in all following illustrations of vectors and matrices in this work.

Here, the two unit processes mining and electricity generation need to be scaled to satisfy the functional unit given in **f**. In a linear process system, the processes are scaled by multiplying the technology matrix **A** by a *scaling vector* **s**:

$$\mathbf{A} \cdot \mathbf{s} = \mathbf{f}. \tag{3.6}$$

The scaling vector contains one scaling factor s_j for each unit process j of the process system. The scaling vector **s** of a process system with a given final demand vector **f** is calculated from Equation 3.6 by inverting the technology matrix **A**, i.e.,

$$\mathbf{s} = \mathbf{A}^{-1} \cdot \mathbf{f} \tag{3.7}$$

for an invertible and positive definite technology matrix **A**.

The resulting elementary flows associated with the process system and a given final demand **f** are obtained from multiplying the intervention matrix **B** by the scaling vector **s**. The resulting total elementary flows are summarized in the *total intervention vector* **g** calculated from

$$\mathbf{g} = \mathbf{B} \cdot \mathbf{s} = \mathbf{B} \cdot \mathbf{A}^{-1} \cdot \mathbf{f}. \tag{3.8}$$

The total intervention vector **g** of the illustrating example is

$$\mathbf{g} = \begin{pmatrix} -372 \\ 1366 \end{pmatrix}. \tag{3.9}$$

Here, producing 1000 kWh electricity requires extracting 372 kg carbon from the environment and causes emissions of 1366 kg CO_2.

Calculating the total intervention vector **g** from Equation 3.8 is the key computational step of LCI-analysis. Therefore, the scaling vector **s** and the total intervention vector **g** are referred to as *LCI-results* in this work.

In LCIA, elementary flows are aggregated to environmental impacts. The aggregation uses *characterization factors* which convert elementary flows to environmental impacts. E.g., characterization factors for the global warming impact (GWI) are global warming potentials. The global warming potential (GWP) measures the contribution of a GHG to global warming relative to the contribution of CO_2(IPCC (2007) on p. 946). The GWP is used to convert elementary flows that contribute to global warming to CO_2-equivalent flows (CO_2-eq.). CO_2-eq. flows are a measure for the environmental impact global warming (IPCC (2007) on pp. 210-216).

The determination of characterization factors for various environmental impacts is an own research field (Margni et al., 2008) and is not further elaborated in this work. The conversion of elementary flows to environmental impacts is computationally realized by multiplying the total intervention vector $\mathbf{g}$ by a *characterization matrix* $\mathbf{Q}$ containing the characterization factors. For the illustrating example, a characterization matrix containing only GWPs is

$$\mathbf{Q} = \begin{pmatrix} 0 & 1 \end{pmatrix}. \tag{3.10}$$

Here, the elementary flow C has a GWP of 0 kg CO_2–eq./kg C and the elementary flow CO_2 has a GWP of 1 kg CO_2–eq./kg CO_2.

The total environmental impacts associated with the process system and a given final demand vector $\mathbf{f}$ are summarized in the *total impact vector* $\mathbf{h}$ calculated from

$$\mathbf{h} = \mathbf{Q} \cdot \mathbf{g} = \mathbf{Q} \cdot \mathbf{B} \cdot \mathbf{A}^{-1} \cdot \mathbf{f}. \tag{3.11}$$

The total GWI h_{GWI} of the illustrating example thus becomes

$$h_{\text{GWI}} = \mathbf{Q} \cdot \mathbf{g} = 1366\,\text{kg CO}_2\text{–eq.}. \tag{3.12}$$

Equation 3.11 represents the computational step of LCIA. Therefore, the total impact vector $\mathbf{h}$ is in this work also referred to as *LCIA-results*. Altogether, the scaling vector $\mathbf{s}$ from Equation 3.7, the total intervention vector $\mathbf{g}$ from Equation 3.8 and the total impact vector $\mathbf{h}$ from Equation 3.11 are summarized as *LCA-results*.

Equation 3.11 is a linear equation system. This linear equation system is well-defined for a square and positive definite technology matrix $\mathbf{A}$. But computational problems arise for over- or under-determined equation systems. In LCA, multi-functional unit processes create an over-determined equation system in Equation 3.11 and cause computational problems. In the following section 3.2, multi-functional unit processes are introduced and the resulting computational complications are discussed.

3.2 Multi-functionality in life cycle assessment

3.2.1 Multi-functional unit processes

In the previous section 3.1.3, the computational structure of LCA is illustrated using two unit processes: mining and electricity generation (Figure 3.4). Each of those two unit processes has only one product (coal and electricity). Therefore, mining and electricity generation are *mono-functional* unit processes. But many industrial processes have multiple products. The electrolysis described in chapter 2, for example, produces chlorine, caustic soda, and the membrane process additionally hydrogen. The electrolysis is therefore a multi-functional unit processes.

A general definition of multi-functionality is based on the functional flows of unit processes. Industrial processes are operated either to manufacture a product or to process a waste. Consequently, the functional flows of industrial processes are either product output flows or waste input flows (section 3.1.2). A unit processes that combines two or more functional flows is *multi-functional*.

Figure 3.5 shows all three possible types of multi-functional unit processes: a) a unit process with multiple product output flows is called *multi-product process*; b) a unit process with multiple waste input flows is named *combined waste processing*; c) a unit process with at least one waste input flow and at least one product output flow is referred to as *recycling process*.

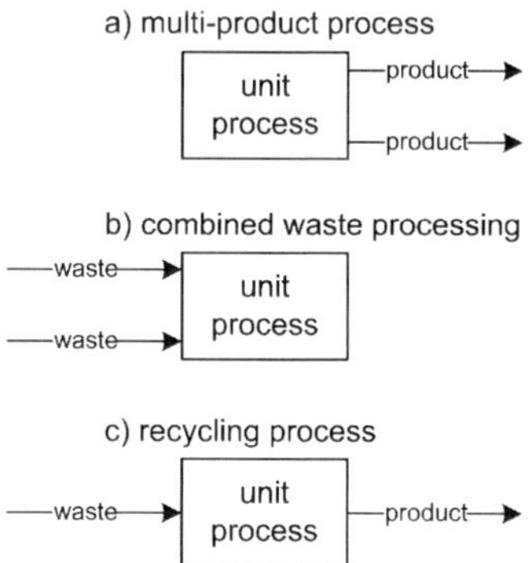

Figure 3.5: Types of multi-functional unit processes: a) multi-product process b) combined waste processing c) recycling process.

The electrolysis processes described in chapter 2 are examples for a multi-product processes. Therefore, all methods and illustrating examples throughout this work employ multi-product processes as well. Still, the methodological problems stated in the subsequent sections 3.2.4 and 3.3.3 exist for all types of multi-functional unit processes. Moreover, the methods developed in the following chapters of this work are also applicable to combined waste processing and recycling processes. The application of these methods to combined waste processing and recycling is briefly covered in the discussion sections of chapter 4, 5, and 6.

All three types of multi-functional unit processes cause a methodological problem. This so-called multi-functionality problem is described and defined in the following section.

3.2.2 The multi-functionality problem in LCA

The environmental impacts of a multi-functional process are shared by the process's functional flows. If a LCA-study does not include all functional flows of a multi-functional process, the LCA-study should also include not all environmental impacts of the multi-functional process. The crucial problem for LCA-practitioners thus is to decide which environmental impacts should be included in a LCA-study. Mathematically, the problem is described by the fact that Equations 3.7, 3.8, or 3.11 cannot be uniquely solved to calculate LCA-results. This problem is called *multi-functionality problem*[5].

However, not every multi-functional unit process causes a multi-functionality problem. Heijungs and Frischknecht (1998) analyzed the mathematical nature of the multi-functionality problem and presented a mathematical definition for the multi-functionality problem. This definition and the mathematical nature of the multi-functionality problem are important fundamentals of this work. Therefore, the definition of Heijungs and Frischknecht (1998) is summarized and briefly illustrated in this section.

The process flow diagram of a process system with a multi-functional unit process is shown in Figure 3.6. The process system includes two unit processes: mining (process 1) and combined heat and power generation (CHP). CHP (process 2) is a multi-product process with two functional flows, i.e., electricity and heat. Process systems containing multi-functional unit processes such as shown in Figure 3.6 are called *multi-functional process systems* in this work.

The functional unit is "producing 1000 kWh electricity from coal". The technology matrix $\mathbf{A}_{\mathrm{mf}}$ and final demand vector $\mathbf{f}_{\mathrm{mf}}$ of the multi-functional process system are

$$\mathbf{A}_{\mathrm{mf}} = \begin{pmatrix} 1 & -0.42 \\ 0 & 1 \\ 0 & 1.67 \end{pmatrix} \text{ and } \mathbf{f}_{\mathrm{mf}} = \begin{pmatrix} 0 \\ 1000 \\ 0 \end{pmatrix}. \tag{3.13}$$

[5]The multi-functionality problem is also referred to as *allocation problem* (Heijungs and Frischknecht, 1998).

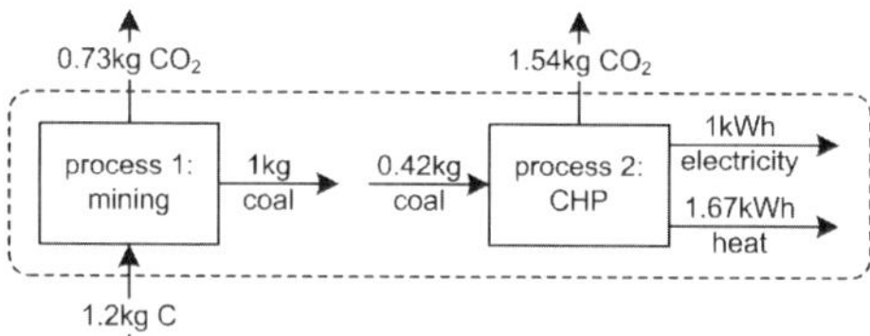

Figure 3.6: Process flow diagram for a process system with two unit processes: mining and combined heat and power generation (CHP). The CHP process is multi-functional because it produces electricity and heat.

Here, the index multi-functional (mf) of $\mathbf{A}$ and $\mathbf{f}$ refers to a multi-functional unit process system to distinguish this example from the earlier discussed process system.

As it can be seen in from the technology matrix $\mathbf{A}_{\text{mf}}$, multi-functional process systems are characterized by a non-square technology matrix with more rows (i.e., economic flows) than columns (i.e., processes). Consequently, the linear equation system given by Equation 3.7 is in general over-determined: it has more equations (one for each economic flow) - than variables (one scaling factors s_j for each process).

For the example above, a scaling vector $\mathbf{s}_{\text{mf}}$ cannot be calculated from Equation 3.7 because the technology matrix $\mathbf{A}_{\text{mf}}$ is non-square and thus not invertible. But invertibility of the technology matrix is not a sufficient criterion for a multi-functionality problem: A scaling vector may be calculated from Equation 3.7 by using a *pseudoinverse* of a non-square technology matrix $\mathbf{A}$.

The pseudoinverse is a generalization of an inverse matrix applicable to non-square matrices (Moore, 1920; Penrose, 1955). The pseudoinverse of the technology matrix $\mathbf{A}$ is denoted as $\mathbf{A}^+$. It can be calculated using the transpose $\mathbf{A}^{\mathrm{T}}$ of the technology matrix $\mathbf{A}$, i.e.,

$$\mathbf{A}^+ = \left(\mathbf{A}^{\mathrm{T}}\mathbf{A}\right)^{-1} \cdot \mathbf{A}^{\mathrm{T}}. \tag{3.14}$$

Equation 3.14 applies to singular and non-square technology matrices (Magnus and Neudecker (2007) on p. 36).[6]

[6]Equation 3.14 refers to the so-called MOORE-PENROSE-inverse independently defined by Moore (1920) and Penrose (1955). The MOORE-PENROSE-inverse is unambiguously defined for matrices with real and complex numbers and therefore used throughout this work (Magnus and Neudecker (2007) on pp. 36-45). More details on pseudoinverses can be found in Ben-Israel and Greville (2003).

Replacing the inverse of $\mathbf{A}$ by its pseudoinverse in Equation 3.7 allows calculating scaling vectors also for non-square technology matrices $\mathbf{A}$, i.e., for multi-functional process systems. But solving the over-determined equation system in Equation 3.7 with the pseudoinverse is an approximation. It is not guaranteed that Equation 3.6 is satisfied, i.e., the product of a non-square technology matrix and a scaling vector calculated with a pseudoinverse is not necessarily equal to the final demand vector $\mathbf{f}$. If Equation 3.6 is satisfied can be examined by calculating a *final supply vector* $\tilde{\mathbf{f}}$ from

$$\tilde{\mathbf{f}} = \mathbf{A} \cdot \mathbf{A}^{+}\mathbf{f}. \tag{3.15}$$

The final supply vector $\tilde{\mathbf{f}}$ describes the set of economic flows actually supplied from a process system. Equation 3.6 is satisfied if the final supply vector $\tilde{\mathbf{f}}$ is equal to the final demand vector $\mathbf{f}$. The discrepancy between final supply vector $\tilde{\mathbf{f}}$ and final demand vector $\mathbf{f}$ is shown in a *discrepancy vector* $\mathbf{d}$ calculated from

$$\mathbf{d} = \tilde{\mathbf{f}} - \mathbf{f} = \mathbf{A} \cdot \mathbf{A}^{+}\mathbf{f} - \mathbf{f}. \tag{3.16}$$

Equation 3.6 is satisfied if the discrepancy vector $\mathbf{d}$ is a zero vector, i.e.,

$$\|\mathbf{d}\| = \left\|\mathbf{A} \cdot \mathbf{A}^{+}\mathbf{f} - \mathbf{f}\right\| = 0. \tag{3.17}$$

Equation 3.17 defines the multi-functionality problem mathematically: a multi-functionality problem exists if the norm of the discrepancy vector $\mathbf{d}$ in Equation 3.17 is nonzero, i.e.,

$$\|\mathbf{d}\| \neq 0. \tag{3.18}$$

This definition applies to multi-functionality problems caused by all types of multi-functional unit processes described in section 3.2.1.

As stated earlier, multi-functional process systems generally cause over-determined equation systems in Equations 3.7, 3.8, and 3.11. If $\|\mathbf{d}\| \neq 0$, the solution of those over-determined equation systems

If $\|\mathbf{d}\| = 0$, no multi-functionality problem exists: The over-determined equation systems given in Equations 3.7, 3.8, or 3.11 contain equations that are linearly depending. Linear dependent equations occur, e.g., if the functional unit includes all functional flows of a multi-functional process.

For the example shown in Figure 3.6, the scaling vector calculated from the pseudoinverse of the technology matrix $\mathbf{A}_{\mathbf{mf}}$ is

$$\mathbf{s}_{\text{mf}} = \mathbf{A}_{\mathbf{mf}}^{+} \cdot \mathbf{f}_{\mathbf{mf}} = \begin{pmatrix} 110.8 \\ 263.9 \end{pmatrix}. \tag{3.19}$$

Here, this scaling vector $\mathbf{s}_{\text{mf}}$ does not satisfy the final demand $\mathbf{f}_{\text{mf}}$ because

$$\|\mathbf{d}_{\text{mf}}\| = \left\|\mathbf{A}_{\text{mf}} \cdot \mathbf{A}_{\text{mf}}^{+}\mathbf{f}_{\text{mf}} - \mathbf{f}_{\text{mf}}\right\| = 857.9. \tag{3.20}$$

This process system thus has a multi-functionality problem for the functional unit "producing 1000 kWh electricity from coal". The following section 3.2.3 reviews methods for fixing multi-functionality problems in LCA.

3.2.3 Fixing multi-functionality problems in LCA

Multi-functionality problems in LCA can be fixed with different methods. Therefore, fixing multi-functionality problems is one of the most controversially discussed aspects in LCA because alternative methods exist (Rebitzer et al., 2004; Reap et al., 2008b; Finnveden et al., 2009; Guinee et al., 2011). The existing methods for fixing multi-functionality problems in LCA are briefly described in the following sections 3.2.3.1, 3.2.3.2, and 3.2.3.3. Existing recommendations for the application of those methods are summarized in section 3.2.3.4. More detailed descriptions of those methods can be found in guiding documents and textbooks on LCA (Baumann and Tillman (2004) on pp. 83ff and pp. 110ff, Guinée et al. (2002) on pp. 505-522, and European Commission (2010b) on pp. 72-81 and pp. 254-272).

If possible, multi-functionality problems can and should be avoided by dividing multi-functional unit processes into sub-processes with individual input and output data (ISO (2006b) in section 4.3.4.2 on p. 29). The methods described in the following sections apply to multi-functional unit processes which cannot be subdivided (e.g., the electrolysis processes described in chapter 2) and for which a multi-functionality problem can thus not be avoided.

3.2.3.1 System expansion

System expansion fixes multi-functionality problems by expanding the functional unit with additional functional flows. These additional functional flows are typically the excess flows from multi-functional unit processes.

System expansion can be applied to the example shown in Figure 3.6 by expanding the functional unit for production of heat. The expanded functional unit is "producing

1000 kWh electricity and 1670 kWh heat from coal". The expanded functional unit is expressed in a final demand vector $\mathbf{f}_{\text{mf,exp}}$ using the index expanded (exp), i.e.,

$$\mathbf{f}_{\text{mf,exp}} = \begin{pmatrix} 0 \\ 1000 \\ 1670 \end{pmatrix}. \tag{3.21}$$

The scaling vector

$$\mathbf{s}_{\text{mf,exp}} = \mathbf{A}_{\text{mf}}^{+} \cdot \mathbf{f}_{\text{mf,exp}} = \begin{pmatrix} 420 \\ 1000 \end{pmatrix} \tag{3.22}$$

now satisfies Equation 3.6 because

$$\left\|\mathbf{d}_{\text{mf,exp}}\right\| = \left\|\mathbf{A}_{\text{mf}} \cdot \mathbf{A}_{\text{mf}}^{+}\mathbf{f}_{\text{mf,exp}} - \mathbf{f}_{\text{mf,exp}}\right\| = 0. \tag{3.23}$$

Here, system expansion successfully fixes the multi-functionality problem by creating two linear depending equations.

System expansion bears two disadvantages which can prevent its application. First, system expansion is limited to LCA-studies where the scope allows expanding the initial functional unit. Many LCA-studies aim at product-specific LCIA-results, i.e., so-called environmental footprints of a single product. Expanding the functional unit for additional functional flows gives LCA-results related to multiple products and is thus not consistent with the scope of product-specific LCIA-results.

The second disadvantage concerns comparative LCA. Comparisons of alternative products or processes should be based on equal functional units (cf. section 3.1.2 and ISO (2006a)). Applying system expansion in comparative LCA might require additional processes (Tillman et al. (1994), Baumann and Tillman (2004) on p. 87). A comparison of the two alternative processes electricity generation (Figure 3.4) and CHP (Figure 3.6) illustrates this dilemma: the electricity generation-process in produces solely electricity. The CHP-process in produces electricity and heat causing a multi-functionality problem for the functional unit "producing 1000 kWh electricity from coal". Fixing this multi-functionality problem with system expansion includes heat in the functional unit. The electricity generation process can only satisfy this expanded functional unit if the process system shown in Figure 3.4 is expanded with an additional process for heat production. Such additional processes used during system expansion are called *added processes* in this work.

Expanded processes influence comparative LCA-results. Therefore, the selection of added processes is crucial for the validity of comparative LCA-results. Unfortunately, it is not elaborated in ISO (2006a) how to select these added processes in comparative LCA (Heijungs and Guinee, 2007). The guides from the UNEP and the EU-commission provide more details on this question. Both recommend that either average *mix processes* or discrete *marginal processes* should be used (Sonnemann and Vigon (2011) on p. 79 and European Commission (2010b) on pp. 88-89).

3.2.3.2 Avoided burden

The avoided burden method for fixing multi-functionality problems, *avoided burden* in short, subtracts avoided environmental impacts from a multi-functional process system. The subtracted impacts account for excess functional flows of a multi-functional unit process that do not correspond to the functional unit. Those functional flows avoid the environmental impacts of an *avoided process*. The avoided process is thus substituted by the multi-functional unit process.[7]

For the illustrating example shown in Figure 3.6, heat is a functional flow of the CHP-process that does not correspond to the functional unit given in Equation 3.13. The CHP-process thus avoids environmental impacts from an avoided process, e.g., a heat generation process. For illustration of the avoided burden approach, a process flow diagram with three unit processes is shown in Figure 3.7. The included unit processes are mining (process 1), CHP (process 2, multi-functional), and heat generation (process 3, avoided process).

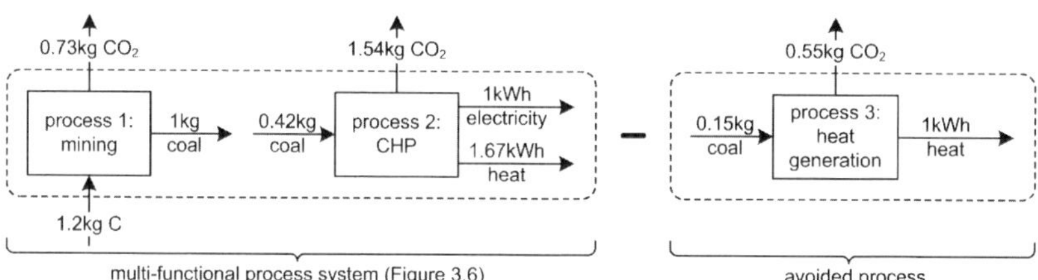

Figure 3.7: Process flow diagram for a process system with three unit processes: mining, CHP, and heat production. Heat production is illustrated as avoided process for heat from CHP.

[7] The avoided burden method is also referred to as substitution method, e.g., by Heijungs and Suh (2002).

The technology matrix $\mathbf{A}_{\text{mf,burden}}$ and final demand vector $\mathbf{f}_{\text{mf,burden}}$ for this multi-functional process system including the avoided process are

$$\mathbf{A}_{\text{mf,burden}} = \begin{pmatrix} 1 & -0.42 & -0.15 \\ 0 & 1 & 0 \\ 0 & 1.67 & 0 \\ 0 & 0 & 1 \end{pmatrix} \text{ and } \mathbf{f}_{\text{mf,burden}} = \begin{pmatrix} 0 \\ 1000 \\ 0 \\ 0 \end{pmatrix}. \tag{3.24}$$

The 4th row in the technology matrix $\mathbf{A}_{\text{mf,burden}}$ and the final demand vector $\mathbf{f}_{\text{mf,burden}}$ represents the heat flow from the heat generation process. This heat flow and the heat flow from the CHP-process (3rd row) are two independent economic flows. The norm of a discrepancy vector calculated from technology matrix $\mathbf{A}_{\text{mf,burden}}$ and final demand vector $\mathbf{f}_{\text{mf,burden}}$ is

$$\left\|\mathbf{d}_{\text{mf,burden}}\right\| = \left\|\mathbf{A}_{\text{mf,burden}} \cdot \mathbf{A}^{+}_{\text{mf,burden}} \mathbf{f}_{\text{mf,burden}} - \mathbf{f}_{\text{mf,burden}}\right\| = 858. \tag{3.25}$$

the multi-functionality problem thus remains.

Applying avoided burden requires assuming that heat from the CHP-process and from the heat generation process are *equivalent economic flows*. Equivalent (eq) economic flows perform the same function. For applying avoided burden, these equivalent economic flows have to be merged into the same row of the technology matrix and the final demand vector, i.e.,

$$\mathbf{A}_{\text{mf,burden,eq}} = \begin{pmatrix} 1 & -0.42 & -0.15 \\ 0 & 1 & 0 \\ 0 & 1.67 & 1 \end{pmatrix} \text{ and } \mathbf{f}_{\text{mf,burden,eq}} = \begin{pmatrix} 0 \\ 1000 \\ 0 \end{pmatrix}. \tag{3.26}$$

A scaling vector $\mathbf{s}_{\text{mf,burden,eq}}$ can then be calculated from Equation 3.7, i.e.,

$$\mathbf{s}_{\text{mf,burden,eq}} = \mathbf{A}^{-1}_{\text{mf,burden,eq}} \cdot \mathbf{f}_{\text{mf,burden,eq}} \begin{pmatrix} 169.5 \\ 1000 \\ -1670 \end{pmatrix}. \tag{3.27}$$

For that scaling vector, Equation 3.17 yields a discrepancy vector with a zero norm.

The negative sign for the third scaling factor in Equation 3.27 corresponds with the sign illustrated in Figure 3.7. It indicates that the environmental impacts of the avoided process are subtracted. Depending on the avoided process, the total impacts of a multi-functional process system can become negative. This is a drawback of using avoided burden because negative impacts are difficult to communicate: they may imply, e.g., that a process system including an avoided burden is absorbing emissions from the environment. In fact, the multi-functional unit process still has emissions.

Avoided burden fixes multi-functionality problems without changing the functional unit. Compared to system expansion, avoided burden can thus also be applied in studies where the functional unit cannot be changed, e.g., environmental footprints of single products. However, this increased flexibility is paid for by always requiring an avoided process. Avoided burden is thus not applicable if no avoided process exists (Curran, 2007b).

The need of added or avoided processes makes system expansion and avoided burden conceptually similar (Azapagic and Clift, 1999a; Heijungs and Guinee, 2007). The selection of avoided processes should follow the same guidelines as for added processes in system expansion, i.e., using mix or marginal processes (Sonnemann and Vigon (2011) on p. 79 and European Commission (2010b) on pp. 88-89).

3.2.3.3 Allocation

Allocation fixes multi-functionality problems by transforming multi-functional unit processes into mono-functional unit processes.[8] As the name indicates, each mono-functional unit process has exactly one functional flow of the multi-functional unit process. The environmental impacts of the multi-functional unit process are allocated to the mono-functional unit processes. For this purpose, all non-functional flows are allocated from the multi-functional unit process to the mono-functional unit processes. Non-functional flows include economic flows and all elementary flows.

For illustration, a process system with allocated mono-functional unit processes is shown in the process flow diagram in Figure 3.8. The process system includes three unit processes: mining (process 1), CHP-electricity (process 2, allocated mono-functional), and CHP-heat (process 3, allocated mono-functional). Here, the mono-functional unit processes origin from the multi-functional CHP-process (process 2 in Figure 3.6). All elementary flows as well as the non-functional economic flows (here the input flow coal) of that multi-functional CHP-process are allocated to the mono-functional unit processes CHP-electricity and CHP-heat using an *allocation factor* C.

[8] Allocation is also referred to as partitioning method, e.g., by Heijungs and Suh (2002).

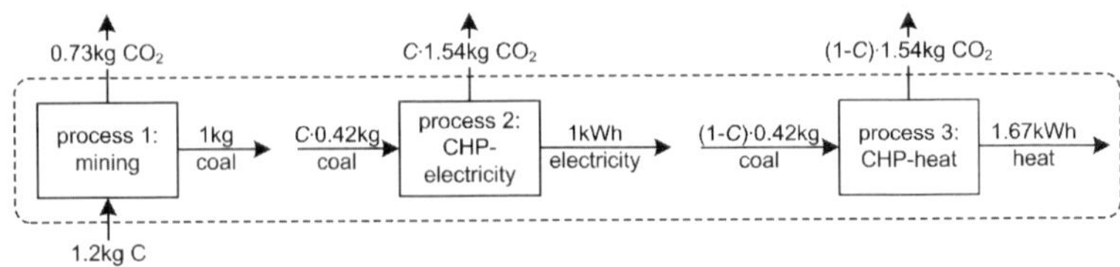

Figure 3.8: Process flow diagram for a process system with three unit processes: mining, CHP-electricity, and CHP-heat. CHP-electricity and CHP-heat are allocated mono-functional unit processes using the allocation factor C.

The allocated technology matrix $\mathbf{A}_{\text{mf,allocated}}$ of the system shown in Figure 3.8 is

$$\mathbf{A}_{\text{mf,allocated}} = \begin{pmatrix} 1 & C \cdot (-0.42) & (1-C) \cdot (-0.42) \\ 0 & 1 & 0 \\ 0 & 0 & 1.67 \end{pmatrix}. \tag{3.28}$$

The scaling vector $\mathbf{s}_{\text{mf,allocated}}$ calculated from the allocated technology matrix $\mathbf{A}_{\text{mf,allocated}}$ depends on the allocation factor C, i.e.,

$$\mathbf{s}_{\text{mf,allocated}} = \begin{pmatrix} C \cdot 420 \\ 1000 \\ 0 \end{pmatrix}. \tag{3.29}$$

Allocation factors are determined by an *allocation criterion*[9]. An allocation criterion is a relationship between input and output flows of a unit process which allow deriving allocation factors.

Allocation criteria should preferably be based on *physical relationships*, i.e., an allocation criterion should "reflect the way in which the inputs and outputs are changed by quantitative changes" in the functional flows (cf. ISO (2006b) in section 4.3.4.2 on p. 29). Such physical relationships can be found if the functional flows of a unit process can be independently varied (Azapagic, 1996; Azapagic and Clift, 1999b,a, 2000; Babusiaux and Pierru, 2007). However, for multi-functional processes with stoichiometrically fixed functional flows, an allocation criterion based on physical relationships cannot be defined. A classic example in LCA literature for stoichiometrically fixed

[9] ISO (2006a) refers to allocation procedures, some publications use the term allocation principle.

functional flows is chlor-alkali electrolysis as described in chapter 2 (e.g., Klöpffer and Grahl (2009) on pp. 98-99, Azapagic and Clift (1999b)).

If physical relationships cannot be determined, allocation criteria should be based on "other relationships" between functional flows and inputs and outputs of a multi-functional unit process (cf. ISO (2006b) in section 4.3.4.2 on p. 29). An example for other relationships are market prices of the economic flows. In so-called *economic allocation*, the allocation factors are based on the market prices of the functional flows. A detailed framework for economic allocation was provided by Frischknecht (1998, 2000). Guinee et al. (2004) added a decision tree for distinguishing economic allocation for multi-product processes, combined waste processing, and recycling. However, the debate on advantages and drawbacks of economic allocation in LCA is remaining open (Ardente and Cellura, 2012; Pelletier and Tyedmers, 2011; Weinzettel, 2012; Pelletier and Tyedmers, 2012). The major criticism of economic allocation is that market prices may fluctuate. Consequently, different and thus uncertain LCA-results are obtained depending on the market prices used for economic allocation.

Besides economic allocation, relationships based on physical properties of the functional flows are widely used as allocation criteria. Common examples for these allocation criteria are masses or energy content of functional flows. Mass and energy contents of a functional flows are typically easy to determine. Still, those allocation criteria are rather arbitrary because they do not necessarily reflect a relationship inputs and outputs of a multi-functional unit process. Therefore, many authors derive allocation criteria which reflect relationships between functional flows and inputs and outputs better than mass or energy content. These allocation criteria usually apply to specific products or groups of products. Some examples for such product-specific allocation criteria are given in Table 3.1.

For most multi-functionality problems, multiple "other relationships" can be applied as allocation criterion. However, ISO (2006b) and other guiding documents lack recommendations for choosing among these criteria.

3.2.3.4 Recommendations for fixing multi-functionality problems in LCA

It is frequently discussed which method LCA-practitioners should preferably use to fix multi-functionality problems in LCA. ISO (2006b) explicitly prefers system expansion over allocation: "wherever possible, allocation should be avoided by [...] expanding the product system to include additional functions" (ISO (2006b) in section 4.3.4.2 on p. 29). If allocation cannot be avoided, allocation criteria should be applied as outlined in section 3.2.3.3 (ISO (2006b) in section 4.3.4.2 on p. 29). The avoided

Table 3.1: Examples and references for product-specific allocation criteria

product	allocation criterion	reference
electricity and heat (CHP)	exergy content	Lucas (2000), Rosen (2008), Aldrich et al. (2011)
	exergetic costs, exergoenvironmental costs	Gonzalez et al. (2003)
dairy products	solid milk content of products	Feitz et al. (2007)
ceramic products	mass, volume, number of pieces	Quinteiro et al. (2012)
window frames	market prices of secondary materials	Werner and Richter (2000)

burden method is not explicitly mentioned in the standard. However, the preference of system expansion is usually interpreted as a preference of system expansion and avoided burden over allocation (Heijungs and Guinee, 2007).

The preference of system expansion and avoided burden over allocation is supported by many publications (e.g., Finnveden and Ekvall (1998); Ekvall (1999a); Weidema (2000); Weidema and Schmidt (2010)). The main argument is that system expansion is capable of integrating the consequences of changes in one product life cycle on another product life cycle. But ISO (2006b) is also often criticized for using vague formulations such as "avoid allocation, if possible" or "other relationships" (Kim and Overcash, 2000; Ekvall and Finnveden, 2001; Heijungs and Guinee, 2007; Curran, 2007b; Reap et al., 2008b). Both Curran (2007b) and Reap et al. (2008a) conclude that more detailed guidance on how to apply methods for fixing multi-functionality problems is required. More precisely, Curran (2007b) demands for "goal-dependent guidance reflecting the user's needs". In recent years, such goal-dependent guidance that is conform to the standard has been developed by organizations such as the UNEP and the EU-commission.

The life cycle initiative of the UNEP published guidance principles for creating LCA databases (Sonnemann and Vigon, 2011). Those authors do not recommend one method over the other but provide practical recommendations how to apply allocation (Sonnemann and Vigon (2011), Figure 3.3 on p. 78) and how to select added or avoided processes for system expansion and avoided burden (Sonnemann and Vigon (2011) on p. 79).

The EU-Commission provided an extensive guide for LCA that conforms to the standard (European Commission, 2010b). This guide includes detailed recommendations for modeling multi-functional process systems and fixing multi-functionality problems (European Commission (2010b) on pp. 72-81 and pp. 254-272). The recommendations give detailed procedures for fixing multi-functionality problems depending on the decision context of a LCA-study (European Commission (2010b) on pp. 36ff) and confirm the standard's preference of system expansion and avoided burden over allocation. The EU-document was developed to provide "the technical basis to derive product-specific criteria, guides, and simplified tools" (European Commission (2010b) on p. ii). In contrast to the standard, both documents from the UNEP and the EU-Commission explicitly mention avoided burden as a variation of system expansion (Sonnemann and Vigon (2011) on p. 79; European Commission (2010b) on pp. 77-78).

In addition the general LCA-guides from the UNEP and the EU-Commission, both British Standards Institution (2011) and World Resources Institute (2010) particularly address product-specific GHG-emissions, i.e., so-called carbon footprint (CFP). In both CFP-guides, avoided burden is also preferred over allocation (British Standards Institution (2011) on p. 22 and World Resources Institute (2010) on p. 66).

Even before the publication of the guides described above, product-specific guides for fixing multi-functionality problems existed, e.g., for wood-based products (Jungmeier et al., 2002a,b), for bioenergy products (Kim and Dale, 2002; Bernesson et al., 2004; Malca and Freire, 2006; Bier et al., 2012; Nguyen and Hermansen, 2012), for pulp and paper (Ekvall, 1999b), for agricultural products (Cederberg and Stadig, 2003). All these publications provide product-specific recommendations for fixing multi-functionality problems in line with ISO (2006b).

For products from chlor-alkali electrolysis, such product-specific guidance does not exist. Moreover, the existing guidance documents focus rather on products than on processes. A guide for fixing multi-functionality problems in line with ISO (2006b) for comparative LCA of multi-functional processes is still lacking. Based on this observation, the following section 3.2.4 states the first methodological research problem discussed in this work.

3.2.4 Problem statement: comparative LCA of multi-product processes

The chlor-alkali electrolysis described in chapter 2 may cause multi-functionality problems in LCA just as any other multi-functional unit process. The methods for fixing those multi-functionality problem are well-known. The existing guides for fixing multi-functionality problems are applicable to LCA-studies of chlor-alkali electrolysis in some particular scopes: a carbon footprint of chlor-alkali electrolysis products may follow British Standards Institution (2011); LCA databases of the processes can be created in line with Sonnemann and Vigon (2011), cf. section 3.2.3.4.

But the scope of comparing multi-product processes with non-common products is little addressed: It is briefly mentioned as "partial equivalence" by (European Commission (2010b) on p. 67). The same authors recommend to "render the systems comparable" by applying methods for fixing multi-functionality problems (European Commission (2010b) on p. 67). In the present thesis, the abstract term partial equivalence is replaced by multi-functionality problem due to non-common products to emphasize the context of comparisons of multi-product processes.

Fixing multi-functionality problems in comparative LCA is not explicitly addressed in the existing literature. Therefore, the aim of this work is to develop a procedure for fixing multi-functionality problems in comparative LCAs of multi-product processes with non-common products. The procedure should conform to the ISO-standard; it should particularly meet the requirement that comparative LCA-studies should use the same functional unit (ISO (2006b) section 4.3.2.7 on p. 22). This requirement raises an issue because comparing multi-product processes with non-common products offers multiple options: The functional unit may in- or exclude the non-common products.

Fixing multi-functionality problems is in many ways linked to the functional unit: system expansion changes the functional unit; depending on the products included in the functional unit, system expansion and avoided burden may require added or avoided processes. A procedure that considers these links between functional unit and fixing multi-functionality problems is therefore developed in chapter 5.

Regardless of preferences between system expansion, avoided burden and allocation, all these methods can introduce uncertainties to LCA-results. The following section reviews these uncertainties and existing methods for analyzing them.

3.3 Analysis of uncertainties due to fixing multi-functionality problems

3.3.1 Uncertainties in LCA

3.3.1.1 Review of uncertainty models in LCA

Uncertainties of scientific models and measurements as well as their quantitative assessment is a widely discussed research field (e.g., Morgan and Henrion (1992)). It has been long recognized that LCA's holistic approach is subject to different uncertainties (Pohl et al., 1996). The categorization of uncertainties in LCA in this work follows several review papers (Huijbregts, 1998a,b, 2001; Björklund, 2002; Heijungs and Huijbregts, 2004; Geisler et al., 2005; Lloyd and Ries, 2007; Baker and Lepech, 2009; Finnveden et al., 2009). Finnveden et al. (2009) distinguish three sources of uncertainty in LCA:

- data or parameters
- choices
- models or relations

These sources are further categorized into uncertainty types. Parameter uncertainty can be due to erroneous, variable, mis-specified, or incomplete data. Choices can be made inconsistently, e.g., in relation with the scope of a study or between alternatives. LCA models may also be wrong (e.g., linear model for non-linear behavior), incomplete or subject to inaccurate software implementations.

Data uncertainties are frequently modeled with probability distributions which represent "the uncertain belief about the likelihood of a parameter having a different possible value" (Morgan and Henrion, 1992). Common probability distributions are uniform, GAUSSIAN-, or lognormal distributions. Probability distributions are typically characterized by the standard deviation σ and the variance σ^2 (Morgan and Henrion, 1992).[10] The standard deviation is the square-root of the variance. Both variance and standard deviation are measures for the variation of a parameter from an average or expected value.

Incorporating probability distributions in LCI-data was discussed by Hanssen and Asbjornsen (1996) and further promoted by, e.g., Maurice et al. (2000) and Sonnemann et al. (2003). The ecoinvent database provided LCI-data with lognormal

[10] The variance is sometimes also referred to as coefficient of variance (CV).

probability distributions for life cycle inventory data (Frischknecht et al., 2005). Still, Heijungs and Frischknecht (2005) describe that LCI-datasets and LCA-software tools often specify probability distributions with different parameters and terminologies. Therefore, uncertainty information are often difficult to transfer from LCI-datasets into software tools for further uncertainty analysis.

An early discussed alternative modeling approach for uncertainties in LCA is fuzzy theory (Chevalier and Tano, 1996; Ros, 1998; Pohl, 1999). The application of fuzzy theory for uncertainty modeling and analysis in LCA has been demonstrated in research-oriented case studies (Mauris et al., 2001; Benetto et al., 2006; Güereca et al., 2007; Seppälä, 2007; Tan, 2008; Heijungs and Tan, 2010; Cruze et al., 2013). Still, fuzzy theory for uncertainty analysis in LCA is less popular than classic probability distributions, mainly because fuzzy theory remains difficult to communicate to non-experts (Lloyd and Ries, 2007) and LCA-software tools are not tailored to handle uncertainty information modeled in fuzzy theory. Therefore, fuzzy theory is not further discussed in this work.

3.3.1.2 Uncertainties due to fixing multi-functionality problems

In section 3.2.3, system expansion, avoided burden, and allocation are discussed as methods for fixing multi-functionality problems in LCA. All these methods introduce uncertainties. The sources of the uncertainties due to fixing multi-functionality problems are choices and use of uncertain data. The different types of uncertainty due to fixing multi-functionality problems are described in this section and summarized in Table 3.2. The last column of the Table 3.2 refers to sections of this work in which methods for modeling and analyzing these uncertainties are presented.

System expansion and avoided burden often require choosing an added or avoided processes between multiple candidate processes (sections 3.2.3.1 and 3.2.3.2). Expanded or avoided processes should be either mix or marginal processes. Both kinds of processes introduce three types of uncertainty in system expansion and avoided burden.

The first type of uncertainty is due to selecting a marginal process as added or avoided process. The source of this uncertainty is a potentially ambiguous choice of the LCA-practitioner (Cooper et al., 2008). Detailed guidelines for determining marginal processes exists, e.g., in European Commission (2010b) on pp. 164-183. But those authors conclude that it is not always possible to determine a single marginal process. In this work, uncertainties due to selecting marginal processes are discussed in section 6.2.2.2.

The choice between selecting one marginal process can also be avoided by using a mix of marginal processes (European Commission (2010b) on pp. 164-183). But mix processes can be subject a second type of uncertainties: The composition of mix processes can vary over time. A good example for a mix process with varying composition is the German electricity mix that changed considerably between 1990 and 2012 due to the phase-out of nuclear power and the increased capacities of renewable power technologies (Arbeitsgemeinschaft Energiebilanzen, 2013). LCA-results are typically calculated using a constant mix composition, e.g., referring to a reference year. Using a mix composition from a past reference year introduces uncertainty to decisions for the future. This uncertainty due to mix processes with constant mix compositions origins in uncertain data, i.e., uncertain mix compositions. Analysis of this uncertainty is discussed in section 6.2.2.1.

When added or avoided processes are required to fix a multi-functionality problem, additional process data from those added or avoided processes is necessary for calculating LCA-results. The additional process data can be uncertain as well; additional process data uncertainty is thus also introduced by fixing multi-functionality problems with system expansion or avoided burden. The third type of uncertainty is therefore referred to as uncertainty due to additional process data and further studied in section 6.2.2.4.

Allocation involves two types of uncertainty. First, uncertainty is introduced by the choice between available allocation criteria and the corresponding allocation factors (section 3.2.3.3). This uncertainty is referred to as uncertainty due to selecting an allocation criterion. A method for analyzing this uncertainty is presented in section 6.2.3.1.

Allocation factors itself can be subject to data uncertainty: Market prices are often uncertain parameter used as criterion for economic allocation. Market prices and the corresponding allocation factor fluctuate over time. But LCA-results are typically calculated for a constant allocation factor. Thereby, economic allocation using a constant allocation factor based on fluctuating market prices introduces uncertainty (Feitz et al., 2007). This type of uncertainty is referred to as uncertainty due to constant allocation factors. It is discussed in section 6.2.3.2.

3.3.2 Existing approaches for uncertainty analysis in LCA

The uncertainty of a model output depends on the input uncertainties and on the sensitivity of the output with respect to the inputs. Sensitivity analysis computes the effects of changes in model inputs on model outputs. A sound uncertainty analysis

Table 3.2: Summary of uncertainties due to fixing multi-functionality problems with system expansion, avoided burden, and allocation; source of these uncertainties; section in this work that covers analysis of the corresponding uncertainties.

method	uncertainty due to...	source	section in this work
system expansion	- selection of marginal process as added process	choice	6.2.2.2
	- added mix process with constant mix composition	data	6.2.2.1
	- additional process data from added process	data	6.2.2.4
avoided burden	- selection of marginal process as avoided process	choice	6.2.2.2
	- avoided mix process with constant mix composition	data	6.2.2.1
	- additional process data from avoided process	data	6.2.2.4
allocation	- selection of allocation criterion	choice	6.2.3.1
	- economic allocation with constant allocation factors	data	6.2.3.2

combines sensitivity analysis with propagation of input uncertainties for "the computation of total uncertainty induced in the output by quantified uncertainty in the inputs and models" (Morgan and Henrion, 1992). The focus of this work is uncertainties in the inputs but not uncertainties in the model.

3.3.2.1 Parameter variation and scenario analysis

Frequently used approaches for uncertainty analysis are parameter variation and scenario analysis. Parameter variation repeatedly calculates results with varying input data. Scenarios denote possible future systems (Börjeson et al., 2006) which are defined by sets of parameters. Börjeson et al. (2006) distinguish six general types of scenarios. From those six types, what-if scenarios are common for comparative LCA (Hojer et al., 2008). What-if scenarios answer the question "what will happen, on the condition of some specified event?" (Börjeson et al., 2006; Hojer et al., 2008).

Scenario analysis compares the results of multiple uncertain scenarios. Scenario analyses are easy to conduct with commercial LCA-software tools (e.g., PE International (2013); ifu (2013); PRé Consultants (2013)). Lloyd and Ries (2007) found scenario analyses to be the second most frequent method for general uncertainty analysis in LCA. For analysis of uncertainties due to fixing multi-functionality problems, it seems that scenario analysis is actually the most frequently applied method: many of the following papers apply scenario analysis in order to quantify the influence of fixing multi-functionality problems on LCA-results.

The influence of varying allocation criteria is often studied for specific product groups such as seafood products (Ayer et al., 2007; Svanes et al., 2011), construction materials (Sayagh et al., 2010; Habert, 2012), fossil fuel chains (Guinee and Heijungs, 2007), or electricity generation (Graus and Worrell, 2011; Soimakallio and Saikku, 2012).

Choosing either allocation or avoided burden is frequently discussed in biofuel research (Bernesson et al., 2004; Malca and Freire, 2006; Curran, 2007a; Guinée et al., 2009; Luo et al., 2009; González-García et al., 2010; Kaufman et al., 2010; Nguyen and Hermansen, 2012; Wardenaar et al., 2013). These authors compare scenarios which apply either allocation or avoided burden.[11]. Varying LCA-results of those scenarios indicate a strong influence of the choice between allocation and avoided burden. Similar scenarios are defined in LCAs of agricultural products, where LCA-results are also found to be strongly influenced by choosing either allocation or avoided burden (Cederberg and Stadig, 2003; Thrane, 2004; Flysö et al., 2011; Bier et al., 2012).

Scenario analysis can suffer from ambiguity if the techniques for generating scenarios are not properly applied (Hojer et al., 2008). The selection of allocation criteria for different scenarios is an example for the potential ambiguity during scenario generation. Spielmann et al. (2005) and Cooper et al. (2008) provided systematic approaches for scenario generation in LCA. Spielmann et al. (2005) derived a method for scenario generation from socio-economic forecasts including a formal consistency check. Cooper et al. (2008) combined the matrix formulation of LCA with decision tree analysis for generating scenarios for process and material alternative. Still, both approaches do not explicitly discuss uncertainties due to fixing multi-functionality problems.

3.3.2.2 Monte-Carlo simulation

Monte-Carlo (MC)-simulation is a frequently used sampling method for uncertainty analysis (Vose, 1996). Based on given input uncertainties, MC-simulations compute a large number of random results (so-called samples). The statistical characteristics of those samples (e.g., probability distribution, variance, standard deviation) provide information on the uncertainty of the actual result.

For LCA, Huijbregts (2001) provided a framework for uncertainty analysis based on

[11] Some of the authors refer to avoided burden as system expansion

MC-simulations. Indeed, many LCAs apply MC-simulations for uncertainty analysis (Lloyd and Ries, 2007). Most frequently, uncertainty of LCI-data is analyzed (e.g., McCleese and LaPuma (2002); Sonnemann et al. (2003); Mattila et al. (2011)). Some studies additionally include uncertainties in LCIA-data, normalization, and weighting data (Huijbregts, 1998a,b; Huijbregts et al., 2003; Hung and Ma, 2009; Mattila et al., 2012).

Uncertainties from fixing multi-functionality problems are only included in some LCAs. Huijbregts (1998b, 2001) combines MC-simulations with scenario analysis to study the influence of the choice between alternative allocation criteria and avoided burden. Similar approaches can be found, e.g., in Geisler et al. (2004), Bojarski et al. (2008), and Maurice et al. (2000). Huijbregts et al. (2003) described how scenario choices due to fixing multi-functionality problems can be systematically integrated to uncertainty analysis. For this purpose, the author combines MC-simulations with parametric bootstrapping. Sills et al. (2013) carried out MC-simulations including probability distributions of alternative avoided processes. In their case study, it turned out that uncertainty of LCA-results using alternative avoided processes are so high that conclusion from single average values can be misleading.

MC-simulations rely on numerical simulations. The number of simulations (or runs) determines the accuracy of the output uncertainty. High accuracy is thus paid for by increased computational effort. Peters (2007) presented an algorithms for more efficient MC-computations of LCA systems. In addition, MC-simulations remain a stochastic method, i.e., results are not exactly reproducible (Heijungs and Suh, 2002).

3.3.2.3 Analytical error propagation

Analytical error propagation can be used for uncertainty analysis if a mathematical relationship between model inputs and outputs is known. The most common analytical approach applies first-order TAYLOR-approximations for modeling uncertainty propagation (Taylor, 1982; Morgan and Henrion, 1992).

For a function $f(x, y)$, the uncertainty of input data x and y can be expressed as standard deviations σ_x and σ_y. The contribution of input uncertainties to an output uncertainty $\sigma(f(x, y))$ is then the product of the output sensitivity with respect to the input and the input uncertainty $\sigma(x)$ (Morgan and Henrion, 1992):

$$\sigma(f(x, y)) = \frac{\partial f(x, y)}{\partial x} \cdot \sigma_x. \tag{3.30}$$

This approach can be used to calculate the total output uncertainty from all input uncertainty contributions. Total output uncertainties are often measured by the output variance, i.e., the sum of the squares of the uncertainty contributions from each input. Using a first-order TAYLOR-approximation for the total variance of $f(x, y)$ and $\mathrm{var}(x) = \sigma_x^2$ yields

$$\mathrm{var}(f) = \left(\frac{\partial f}{\partial x}\right)^2 \cdot \mathrm{var}(x) + \left(\frac{\partial f}{\partial y}\right)^2 \cdot \mathrm{var}(y) + 2\frac{\partial f}{\partial x}\frac{\partial f}{\partial y}\mathrm{cov}(x, y), \tag{3.31}$$

where $\mathrm{cov}(x, y)$ is the covariance between x and y. The partial derivatives in Equation 3.31 are called sensitivity coefficients. Sensitivity coefficients can be used for sensitivity analysis.

This analytical approach was applied to both the sequential and the matrix approach in LCA (section 3.1.3). Analytical formulae for error propagation of sequentially modeled LCA were presented by Ciroth (2001) and Ciroth et al. (2004). Heijungs (1994, 1996); Heijungs and Suh (2002) and Sakai and Yokoyama (2002) applied analytical error propagation for sensitivity and uncertainty analysis in matrix-based LCA. A computational framework for analytical error propagation for analysis of data uncertainties in LCA is fully worked out and promises lower computational effort than MC-simulations (Heijungs, 2010). Still, examples for application of analytical error propagation in LCA are rare, likely because most LCA software does not feature the method yet.[12]

It was recognized that analytical error propagation could in principle be applied to allocation factors as well (Heijungs and Kleijn, 2001; Heijungs and Suh, 2002). Uncertainties from fixing multi-functionality problems have not explicitly been considered in the analytical approach. Indeed, Heijungs (2010) assumes a multi-functionality problem to be already fixed by using a square technology matrix.

3.3.3 Problem statement: uncertainty analysis for fixing multi-functionality problems

Uncertainties from fixing multi-functionality problems are rarely systematically analyzed. The choice between allocation criteria is mostly studied using parameter variation and scenario analysis. MC-simulations sporadically include uncertainties due to choices between allocation criteria in LCA. In both approaches, the contribution of uncertainties due to choices during fixing multi-functionality problems is rarely

[12] The software CMLCA (Heijungs, 2013a) is an exception where the formulas provided by Heijungs (2010) are implemented.

compared to uncertainties in process data. Moreover, the uncertainties illustrated in Table 3.2 are often not integrated in an uncertainty analysis, even though the ISO standard demands that "whenever several alternative allocation procedures seem applicable, a sensitivity analysis shall be conducted to illustrate the consequences of the departure from the selected approach" (ISO (2006b) on p. 28).

A systematic method for conducting that sensitivity analysis, however, is so far lacking. Analytical error propagation seems a promising approach for a holistic uncertainty analysis method: The equations of Heijungs (2010) allow a comparison of different sources of uncertainty in LCA. But the analytical approach does not include uncertainties due to fixing multi-functionality problems yet.

This work aims to develop an uncertainty analysis method including all uncertainties due to fixing multi-functionality problems shown in Table 3.2. For this purpose, the matrix framework (section 3.1.3.1) is expanded for equations modeling system expansion, avoided burden, and allocation in the following chapter 4. Based on the expanded matrix formulation, a holistic method for uncertainty analysis is derived in chapter 6. This method includes all uncertainties due to fixing multi-functionality problems in LCA. The method is applied to study uncertainties from fixing multi-functionality problems in a comparative LCA of chlor-alkali electrolysis in chapter 7.

Chapter 4

A matrix formulation for fixing multi-functionality problems in LCA

Analytical uncertainty analysis requires a functional relationship between the input and output parameters of an LCA model (sections 3.3.2.3 and 3.3.3). The matrix formulation presented in section 3.1.3 is such a functional relationship. However, the methods for fixing multi-functionality problems are not explicitly implemented in that matrix formulation.[1] Consequently, existing analytical uncertainty analysis does not yet include uncertainties due to fixing multi-functionality problems. In order to fill this gap, matrix-based formulations for fixing multi-functionality problems are derived in this chapter.

The formulation expands the existing formulas from Heijungs and Suh (2002) described in section 3.1.3. The new expanded formulas serve two purposes: they provide the functional relationship required for analytical uncertainty analysis and they introduce parameters to represent uncertainties due to fixing multi-functionality problems (cf. Table 3.2). The formulas are also included in Jung et al. (2012, 2013b), the latter was submitted for publication at the time of writing.

The presented matrix formulation provides formulas for system expansion, avoided burden, and allocation. System expansion and avoided burden lead to the same

[1]In the matrix-based software tool CMLCA (Heijungs, 2013a), the methods for fixing multi-functionality problem are not implemented by explicit formulas. Instead, "if-then" constructions are used (Heijungs, 2013b). The calculation procedures of Simapro (PRé Consultants, 2013) are not publicly available.

matrix-based formulas which are presented in section 4.1. Formulas for allocation are discussed in section 4.2. In section 4.3, these formulas are combined into a set of equation that represents an expanded matrix formulation for LCA. The chapter concludes with a discussion of that expanded matrix formulation in section 4.4.

The formulations in the following sections are given for a process system described by a $(m \times n)$-technology matrix $\mathbf{A}$ and a corresponding $(v \times n)$-intervention matrix $\mathbf{B}$:

$$\mathbf{A} = \begin{pmatrix} a_{11} & \cdots & a_{1n} \\ \vdots & \ddots & \vdots \\ \vdots & \ddots & \vdots \\ a_{m1} & \cdots & a_{mn} \end{pmatrix} \text{ and } \mathbf{B} = \begin{pmatrix} b_{11} & \cdots & b_{1n} \\ \vdots & \ddots & \vdots \\ b_{v1} & \cdots & b_{vn} \end{pmatrix}. \tag{4.1}$$

The process system includes at least one multi-functional unit process, i.e., $m > n$. The functional unit is represented by a final demand vector $\mathbf{f}$. A $(u \times v)$-characterization matrix $\mathbf{Q}$ completes the given data:

$$\mathbf{f} = \begin{pmatrix} f_1 \\ \vdots \\ f_m \end{pmatrix} \text{ and } \mathbf{Q} = \begin{pmatrix} q_{11} & \cdots & q_{1v} \\ \vdots & \ddots & \vdots \\ q_{u1} & \cdots & q_{uv} \end{pmatrix}. \tag{4.2}$$

The formulations derived in this chapter are illustrated with a process system illustrated in the process flow diagram shown in Figure 4.1. The process system contains four unit processes: mining (process 1), CHP (process 2), old boiler (process 3), and new boiler (process 4). The old boiler and new boiler produce heat that is equivalent to the heat from the CHP-process.

The corresponding technology and intervention matrices $\mathbf{A}_{\text{CHP}}$ and $\mathbf{B}_{\text{CHP}}$ are

$$\mathbf{A}_{\text{CHP}} = \begin{pmatrix} 1 & -0.42 & -0.15 & -0.13 \\ 0 & 1 & 0 & 0 \\ 0 & 1.67 & 0 & 0 \\ 0 & 0 & 1 & 0 \\ 0 & 0 & 0 & 1 \end{pmatrix} \tag{4.3}$$

and

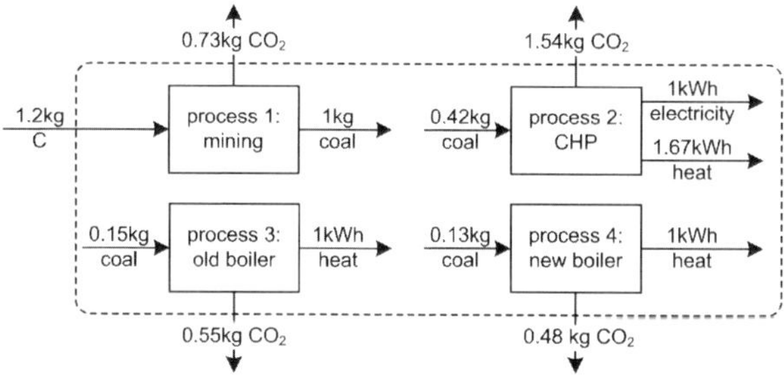

Figure 4.1: Process flow diagram of a process system with four unit processes: mining, CHP, old boiler, and new boiler.

$$\mathbf{B}_{\mathrm{CHP}} = \begin{pmatrix} -1.2 & 0 & 0 & 0 \\ 0.73 & 1.54 & 0.55 & 0.48 \end{pmatrix}. \tag{4.4}$$

The functional unit is "producing 1000 kWh electricity" expressed in a final demand vector $\mathbf{f}_{\mathrm{CHP}}$, i.e.,

$$\mathbf{f}_{\mathrm{CHP}} = \begin{pmatrix} 0 \\ 1000 \\ 0 \\ 0 \\ 0 \end{pmatrix} \tag{4.5}$$

Finally, a hypothetical impact matrix $\mathbf{Q}_{\mathrm{CHP}}$ concludes the example, i.e.,

$$\mathbf{Q}_{\mathrm{CHP}} = \begin{pmatrix} -1 & 0 \\ 0 & 1 \end{pmatrix}. \tag{4.6}$$

4.1 Matrix formulation for avoided burden and system expansion

Both system expansion and avoided burden require selecting an added or avoided process (sections 3.2.3.1, 3.2.3.2). The choice of selecting an added or avoided process can be represented in the same matrix formulation. The formulation is elaborated for the avoided burden method in the following sections 4.1.1 and 4.1.2, because this approach does not change the initial functional unit defined in Equation 4.5. In addition to the selection of an added process, system expansion requires to change the initial functional unit. A matrix formulation for system expansion that accounts for a change in the initial functional unit is therefore discussed in section 4.1.3.

4.1.1 Merging equivalent economic flows

Avoided burden fixes multi-functionality problems by subtracting avoided environmental impacts from a multi-functional process system (section 3.2.3.2). Calculating LCA-results with avoided burden requires merging equivalent economic flows into the same row in the technology matrix $\mathbf{A}$ and the final demand vector $\mathbf{f}$, cf. Equation 3.26 on p. 30. In some cases, process vectors are already specified with equivalent economic flows merged into the same row, e.g., electricity from a power plant is in the same row as electricity from waste incinerator. In this work, however, it is assumed that equivalent economic flows still have to be merged into the same row in the technology matrix $\mathbf{A}$ and the final demand vector $\mathbf{f}$ when avoided burden is applied.

Moreover, it is assumed that equivalent economic flows are specified in the same unit. The assumption is justified because equivalent economic flows with differing units or flow qualities can be easily reformulated to have identical units (cf. Heijungs and Suh (2002) on p. 45).

Merging is realized by adding up rows of equivalent economic flows in the technology matrix $\mathbf{A}$ and the final demand vector $\mathbf{f}$. Rows of the technology matrix are added up by a left matrix multiplication with a $(r \times m)$-*equivalence matrix* $\mathbf{E}$:

$$\mathbf{E} = \begin{pmatrix} e_{11} & \cdots & \cdots & e_{1m} \\ \vdots & \ddots & \ddots & \vdots \\ e_{r1} & \cdots & \cdots & e_{rm} \end{pmatrix}. \tag{4.7}$$

This equivalence matrix is constructed from an initial $(m \times m)$-identity matrix. If $(\alpha + 1)$ economic flows are equivalent, and these equivalent flows are in the kth to $(k + \alpha)$th row of the technology matrix $\mathbf{A}$, then the $(k + 1)$th to $(k + \alpha)$th rows of the initial identity matrix are deleted. The matrix elements $e_{k(k+1)}$ to $e_{k(k+\alpha)}$ are set $e_{k(k+1)} = \cdots = e_{k(k+\alpha)} = 1$.

For the illustrating example, three heat flows from the CHP-process and the old boiler process and the new boiler processes are assumed to be equivalent, i.e., $\alpha = 2$. The heat flows are the 3rd, 4th, and 5th flow in the technology matrix $\mathbf{A}_{\text{CHP}}$. The initial equivalence matrix $\mathbf{E}_{\text{CHP,ini}}$ is a (5×5)-identity matrix. Here, the index refers to initial (ini). Deleting the 4th and 5th row in $\mathbf{E}_{\text{CHP,ini}}$ and setting $e_{34} = e_{35} = 1$

yields the equivalence matrix $\mathbf{E}_{\text{CHP}}$, i.e.,

$$\mathbf{E}_{\text{CHP,ini}} = \begin{pmatrix} 1 & 0 & 0 & 0 & 0 \\ 0 & 1 & 0 & 0 & 0 \\ 0 & 0 & 1 & \mathbf{0} & \mathbf{0} \\ 0 & 0 & 0 & 1 & 0 \\ 0 & 0 & 0 & 0 & 1 \end{pmatrix} \rightarrow \mathbf{E}_{\text{CHP}} = \begin{pmatrix} 1 & 0 & 0 & 0 & 0 \\ 0 & 1 & 0 & 0 & 0 \\ 0 & 0 & 1 & \mathbf{1} & \mathbf{1} \end{pmatrix}. \tag{4.8}$$

The left matrix multiplication by the equivalence matrix is applied to both technology matrix $\mathbf{A}$ and the final demand vector $\mathbf{f}$. For the illustrating example, multiplying $\mathbf{A}_{\text{CHP}}$ and $\mathbf{f}_{\text{CHP}}$ by the equivalence matrix $\mathbf{E}_{\text{CHP}}$ yields

$$\mathbf{E}_{\text{CHP}} \cdot \mathbf{A}_{\text{CHP}} = \begin{pmatrix} 1 & -0.42 & -0.15 & -0.13 \\ 0 & 1 & 0 & 0 \\ 0 & 1.67 & \mathbf{1} & \mathbf{1} \end{pmatrix} \tag{4.9}$$

and

$$\mathbf{E}_{\text{CHP}} \cdot \mathbf{f}_{\text{CHP}} = \begin{pmatrix} 0 \\ 1000 \\ \mathbf{0} \end{pmatrix}. \tag{4.10}$$

This example here has two candidate processes for an avoided process, i.e., the old and new boiler processes (Figure 4.1). The heat flows from the old and new boiler processes are both equivalent to the heat flow of the CHP-process. Calculating a scaling vector with $\mathbf{EA}_{\text{CHP}}$ and $\mathbf{Ef}_{\text{CHP}}$ from Equation 3.7 still yields a non-zero discrepancy vector, cf. Equation 3.16.

In contrast to the example in section 3.2.3.2 (Equation 3.26), merging more than two equivalent economic flows turns the over-determined equation system described by $\mathbf{s} = \mathbf{A}_{\text{CHP}}^{-1} \cdot \mathbf{f}_{CHP}$ into an under-determined equation system described by $\mathbf{s} = (\mathbf{E}_{\text{CHP}}\mathbf{A}_{\text{CHP}})^{-1} \cdot \mathbf{E}_{\text{CHP}}\mathbf{f}_{\text{CHP}}$. An under-determined equation system in Equation 3.7 is always formed if more than $(m-n)$ equivalent economic flows are merged into the same rows, i.e., $(\alpha+1) > (m-n)$. By implication, avoided burden only fixes the multi-functionality problem if exactly two equivalent economic flows are merged; and two economic flows are merged if exactly one avoided process is used.

The under-determined equation system could thus be avoided by merging exactly two economic flows. LCA-practitioners typically use exactly one avoided process and thus merge exactly two economic flows. Thereby, the LCA-practitioner always comes up with a well-defined equation system as shown in section 3.2.3.2 (Equation 3.26). However, they might have to select this one avoided process from several candidates. This selection is one source of uncertainty due to fixing multi-functionality problems (cf. Table 3.2). The uncertainty of selecting from several avoided processes is thus represented by the under-determined equation system described above.

This work aims to analyze the uncertainties due to fixing multi-functionality problems. To analyze the uncertainty due to selecting from multiple avoided processes, the proposed matrix formulation allows merging more than two economic flows, i.e., multiple avoided processes. To still arrive at a well-defined equation system, these avoided processes are aggregated into exactly one aggregated avoided process. Thereby, the parameters modelling the aggregation can be used to analyze the uncertainty from selecting between multiple avoided processes.

The concept of an aggregated unit process allows also to model mix unit process such as an electricity mix. In this case, the parameters modelling the aggregation can also be used to analyze the uncertainty in mix processes with a constant mix composition (cf. Table 3.2). In the following section 4.1.2, a matrix formulation for aggregating multiple avoided processes is presented.

4.1.2 Aggregating avoided processes

Aggregating multiple avoided processes into exactly one avoided process overcomes an under-determined equation system to finally fix a multi-functionality problem. A total number of α unit processes can be aggregated using *aggregation factors* v_j. The process vector $\tilde{\mathbf{p}}$ of the aggregated unit process is calculated from α process vectors $\mathbf{p}_j$ and aggregation factors v_j, i.e.,

$$\tilde{\mathbf{p}} = \sum_{j=1}^{\alpha} v_j \mathbf{p}_j. \tag{4.11}$$

Theoretically, all n unit processes can be used for aggregation. To emphasize that LCA-practitioners typically choose from a subset of all n unit processes, a total number of $\alpha \leq n$ unit processes are used for aggregation in this work. The environmental impacts of the single aggregated process should be equal to the sum of weighted impacts from the original processes. This condition is met if the aggregation factors v_j sum up to one, i.e.,

$$\sum_{j=1}^{n} v_j = 1 \tag{4.12}$$

The aggregation factors are the additional variables required to turn the underdetermined equation system into a well-defined equation system. Moreover, the aggregation factors are continuous parameters that allow applying analytical methods for uncertainty analysis.

Matrix-based aggregation of multiple avoided processes is realized with a right matrix multiplication of the technology matrix $\mathbf{A}$ by a $(n \times w)$-*aggregation matrix* $\mathbf{V}$ that contains aggregation factors v_{jl}:

$$\mathbf{V} = \begin{pmatrix} v_{11} & \cdots & v_{1w} \\ \vdots & \ddots & \vdots \\ \vdots & \ddots & \vdots \\ v_{n1} & \cdots & v_{nw} \end{pmatrix}. \tag{4.13}$$

The aggregation matrix is constructed from an initial $(n \times n)$-identity matrix. If a total number of α avoided processes are represented by the lth to $(l+\alpha-1)$th column of the technology matrix $\mathbf{A}$, then the $(l+1)$ to $(l+\alpha-1)$th columns of the initial identity matrix are deleted. The matrix elements v_{ll} to $v_{(l+\alpha)l}$ are aggregation factors of the lth to $(l+\alpha)$th avoided processes, respectively.

In the example, both the old boiler and the new boiler process are avoided processes, i.e., $\alpha = 2$. The avoided processes are given by the 3rd and 4th columns of the technology matrix $\mathbf{A}_{\text{CHP}}$. Thus, deleting the 4th column in an initial (4×4)-identity matrix $\mathbf{V}_{\text{CHP,ini}}$ gives the aggregation matrix $\mathbf{V}_{\text{CHP}}$ with the aggregation factors $v_{\text{oldBoiler}}$ and $v_{\text{newBoiler}}$, i.e.,

$$\mathbf{V}_{\text{CHP,ini}} = \begin{pmatrix} 1 & 0 & 0 & \not{0} \\ 0 & 1 & 0 & \not{0} \\ 0 & 0 & 1 & \not{0} \\ 0 & 0 & 0 & \not{1} \end{pmatrix} \rightarrow \mathbf{V}_{\text{CHP}} = \begin{pmatrix} 1 & 0 & 0 \\ 0 & 1 & 0 \\ 0 & 0 & v_{\text{oldBoiler}} \\ 0 & 0 & v_{\text{newBoiler}} \end{pmatrix}. \tag{4.14}$$

Purely for illustration, it is assumed that the aggregated avoided process is composed from $v_{\text{oldBoiler}} = 0.7$ of the 3rd process and $v_{\text{newBoiler}} = 0.3$ of the 4th process. Here, a left matrix multiplication of the technology matrix $\mathbf{A}_{\text{CHP}}$ by the equivalence matrix $\mathbf{E}_{\text{CHP}}$ combined with a right matrix multiplication by the aggregation matrix $\mathbf{V}_{\text{CHP}}$ yields

$$\mathbf{E}_{\text{CHP}} \cdot \mathbf{A}_{\text{CHP}} \cdot \mathbf{V}_{\text{CHP}} = \begin{pmatrix} 1 & -0.42 & -0.144 \\ 0 & 1 & 0 \\ 0 & 1.67 & 1 \end{pmatrix}. \tag{4.15}$$

After aggregation of multiple avoided processes, the scaling vector $\mathbf{s}$ is calculated from

$$\mathbf{s} = (\mathbf{EAV})^{-1} \cdot \mathbf{Ef}. \tag{4.16}$$

Aggregating unit processes also affects the intervention matrix $\mathbf{B}$ because that matrix contains the elementary flows of the avoided processes. The intervention matrix $\mathbf{B}$ is therefore also multiplied by the aggregation matrix $\mathbf{V}$. The total intervention vector $\mathbf{g}$ and the total impact vector $\mathbf{h}$ are thus calculated from

$$\mathbf{g} = \mathbf{BV} \cdot (\mathbf{EAV})^{-1} \cdot \mathbf{Ef}, \tag{4.17}$$
$$\mathbf{h} = \mathbf{Q} \cdot \underbrace{\mathbf{BV}}_{\mathbf{B}_{\text{burden}}} \cdot \underbrace{(\mathbf{EAV})}_{\mathbf{A}_{\text{burden}}}{}^{-1} \cdot \underbrace{\mathbf{Ef}}_{\mathbf{f}_{\text{burden}}}. \tag{4.18}$$

The subscripted braces in Equation 4.18 illustrate that the new formulas have the same structure as the original matrix formulation provided by Heijungs and Suh (2002) (cf. Equations 3.7, 3.8, and 3.11). The original technology matrix $\mathbf{A}$ is multiplied by the equivalence matrix $\mathbf{E}$ and the aggregation matrix $\mathbf{V}$. The matrix product $\mathbf{EAV}$ can be seen as an avoided burden-modified technology matrix $\mathbf{A}_{\text{burden}}$. Similarly, the matrix products $\mathbf{BV}$ and $\mathbf{Ef}$ represent a modified intervention matrix $\mathbf{B}_{\text{burden}}$ and a modified final demand vector $\mathbf{f}_{\text{burden}}$, respectively. The new formulas can in fact be reduced to the original matrix formulation in case that $\mathbf{E}$, $\mathbf{A}$ and $\mathbf{V}$ are square matrices and $\mathbf{E}$ and $\mathbf{V}$ are identity matrices because $(\mathbf{EAV})^{-1} = \mathbf{E}^{-1} \cdot \mathbf{A}^{-1} \cdot \mathbf{V}^{-1}$.

In Equations 4.16, 4.17, and 4.18, the key computational step remains a matrix inversion: Here, the avoided burden-modified technology matrix $\mathbf{EAV}$ has to be inverted. Merging economic flows and aggregating unit processes decreases the number of rows and columns. Consequently, the dimension of the matrix $\mathbf{EAV}$ is in fact smaller than the dimension of the original technology matrix $\mathbf{A}$. Therefore, the computational effort for calculating LCA-results is not expected to increase for the formulas in Equations 4.16, 4.17, and 4.18.

4.1.3 Matrix-formulation for system expansion

For system expansion, merging of equivalent economic flows and aggregating process is only required in comparative LCAs[2]. But system expansion also requires changing the functional unit expressed in the final demand vector $\mathbf{f}$. System expansion is not fully described by the formulas presented in sections 4.1.1 and 4.1.2 because the change of the functional unit is not represented. In this section, a matrix formulation is given for changing the functional unit during system expansion.

System expansion changes the final demand vector so that the discrepancy vector has a zero-norm to fix a multi-functionality problem, cf. Equation 3.18. This modification is explained for the illustrating example in the following. The initial final demand vector of the illustrating example includes only the functional flow electricity (Equation 4.5). To illustrate the mathematical idea behind system expansion, the second functional flow of the CHP-process (i.e., heat) is kept as a variable φ in the third element of the final demand vector $\mathbf{f}_{\mathrm{CHP}}$, i.e.,

$$\mathbf{f}_{\mathrm{CHP}} = \begin{pmatrix} 0 \\ 1000 \\ \varphi \\ 0 \\ 0 \end{pmatrix} \tag{4.19}$$

The mathematical idea behind system expansion is to choose the parameter φ so that the discrepancy vector in Equation 3.16 has a zero norm. This is accomplished in the illustrating example by using the exact figure of the heat flow from the CHP-process also for the variable φ, i.e.,

$$\mathbf{f}_{\mathrm{CHP,expanded}} = \begin{pmatrix} 0 \\ 1000 \\ 1670 \\ 0 \\ 0 \end{pmatrix}. \tag{4.20}$$

LCA-results can be calculated from Equations 4.16, 4.17, and 4.18 with a discrepancy vector that has a zero norm.

[2] Heijungs and Suh (2002) show that system expansion can be applied without merging equivalent flows: These authors formulate a linear programming case so that a process system produces at least the products of the functional unit. In this work, only process system producing exactly the products of the functional unit are considered. Therefore, the linear programming approach is not further discussed here.

In a comparative LCA, the expanded functional unit is satisfied by several alternative processes. The functional unit is then represented by one final demand vector for each alternative process. If an alternative process produces only electricity, it becomes necessary to select an added process for heat production. Here, LCA-practitioners might have to select between multiple candidates for added processes. This choice is conceptually equivalent to selecting between multiple avoided processes (section 3.3.1.2). Therefore, the matrix formulas for aggregating multiple avoided processes presented in sections 4.1.1 and 4.1.2 also apply to aggregating multiple added processes for system expansion in comparative LCA.

4.2 Matrix formulation for allocation

Allocation divides multi-functional unit processes into mono-functional unit processes. Each mono-functional process has exactly one functional flow of the multi-functional process. The non-functional flows of the multi-functional unit process are allocated to these mono-functional unit processes, cf. section 3.2.3.3. This procedure cannot be formulated as a single matrix multiplication.

Instead, the matrix formulation described in the following section 4.2.1 first divides multi-functional processes into mono-functional unit processes. Matrix-based allocation of the non-functional flows from the multi-functional to the mono-functional unit processes is then introduced in section 4.2.2. For the ease of explanation, it is assumed that only one multi-functional process is treated by allocation. However, the procedures described in the following sections 4.2.1 and 4.2.2 can be repeated as often as necessary for real process systems with several multi-functional processes.

4.2.1 Transforming multi-functional into mono-functional unit processes

Multi-functional unit processes are transformed into mono-functional processes in two steps. First, the column of a multi-functional unit process is duplicated β times in both the technology matrix $\mathbf{A}$ and the intervention matrix $\mathbf{B}$. Here, β is the number of functional flows of a multi-functional unit process. In the second step, the β duplicated columns are turned into mono-functional unit processes. For this purpose, all economic and elementary flows have to be set to zero except for one functional flow in each duplicated column.

The two transformation steps are realized by multiplying the technology matrix $\mathbf{A}$

by a *transformation matrix* $\mathbf{T}_1$ and the intervention matrix $\mathbf{B}$ by a transformation matrix $\mathbf{T}_0$. Two transformation matrices $\mathbf{T}_1$ and $\mathbf{T}_0$ are required because the technology matrix $\mathbf{A}$ contains functional and non-functional flows while the intervention matrix $\mathbf{B}$ contains only non-functional flows. However, the dimension of both transformation matrices $\mathbf{T}_1$ and $\mathbf{T}_0$ is $n \times (n+\beta)$, i.e.,

$$\mathbf{T}_1 = \begin{pmatrix} t_{11} & \cdots & \cdots & t_{1(n+\beta)} \\ \vdots & \ddots & \ddots & \vdots \\ t_{n1} & \cdots & \cdots & t_{n(n+\beta)} \end{pmatrix} \tag{4.21}$$

and

$$\mathbf{T}_0 = \begin{pmatrix} t_{11} & \cdots & \cdots & t_{1(n+\beta)} \\ \vdots & \ddots & \ddots & \vdots \\ t_{n1} & \cdots & \cdots & t_{n(n+\beta)} \end{pmatrix} \tag{4.22}$$

(4.23)

Both transformation matrices $\mathbf{T}_1$ and $\mathbf{T}_0$ are constructed from an initial $n \times n$-identity matrix. If the lth process in the technology matrix is multi-functional with β functional flows, β new columns of row dimension n are inserted to the initial $n \times n$-identity matrices at positions $(l+1)\ldots(l+\beta)$.

For the transformation matrix $\mathbf{T}_1$ of the technology matrix $\mathbf{A}$, the $(l+1)$th to $(l+\beta)$th inserted columns contain zeroes except for the lth row. The elements in the lth row of all inserted columns are $t_{l(l+1)} = \cdots = t_{l(l+\beta)} = 1$.

In the illustrating example, the multi-functional CHP-process is in the $l = 2$ column of the technology matrix $\mathbf{A}_{\text{CHP}}$ and has $\beta = 2$ functional flows. Thus, $\beta = 2$ columns are added to the initial (4×4) identity matrix $\mathbf{T}_{1,\text{CHP,ini}}$. All elements of those inserted columns are zero except for those in the $l =$ 2nd row. The elements t^1_{23} and t^1_{24} are $t^1_{23} = t^1_{24} = 1$ so that the transformation matrix $\mathbf{T}_{1,\text{CHP}}$ becomes

$$\mathbf{T}_{1,\text{CHP,ini}} = \begin{pmatrix} 1 & 0 & 0 & 0 \\ 0 & 1 & 0 & 0 \\ 0 & 0 & 1 & 0 \\ 0 & 0 & 0 & 1 \end{pmatrix} \rightarrow \mathbf{T}_{1,\text{CHP}} = \begin{pmatrix} 1 & 0 & 0 & 0 & 0 & 0 \\ 0 & 1 & 1 & 1 & 0 & 0 \\ 0 & 0 & 0 & 0 & 1 & 0 \\ 0 & 0 & 0 & 0 & 0 & 1 \end{pmatrix}. \tag{4.24}$$

In contrast, the transformation matrix $\mathbf{T}_0$ of the intervention matrix $\mathbf{B}$ contains only zeroes in the $(l+1)$th to $(l+\beta)$th added columns. In the illustrating example, the transformation matrix $\mathbf{T}_{0,\text{CHP}}$ is

$$\mathbf{T}_{0,\text{CHP,ini}} = \begin{pmatrix} 1 & 0 & 0 & 0 \\ 0 & 1 & 0 & 0 \\ 0 & 0 & 1 & 0 \\ 0 & 0 & 0 & 1 \end{pmatrix} \rightarrow \mathbf{T}_{0,\text{CHP}} = \begin{pmatrix} 1 & 0 & 0 & 0 & 0 & 0 \\ 0 & 1 & 0 & 0 & 0 & 0 \\ 0 & 0 & 0 & 0 & 1 & 0 \\ 0 & 0 & 0 & 0 & 0 & 1 \end{pmatrix}. \tag{4.25}$$

Multiplying the technology matrix $\mathbf{A}_{\text{CHP}}$ by the transformation matrix $\mathbf{T}_{1,\text{CHP}}$ and the intervention matrix $\mathbf{B}_{\text{CHP}}$ by the transformation matrix $\mathbf{T}_{0,\text{CHP}}$ yields

$$\mathbf{A}_{\text{CHP}} \cdot \mathbf{T}_{1,\text{CHP}} = \begin{pmatrix} 1 & -0.42 & -0.42 & -0.42 & -0.15 & -0.13 \\ 0 & 1 & 1 & 1 & 0 & 0 \\ 0 & 1.67 & 1.67 & 1.67 & 0 & 0 \\ 0 & 0 & 0 & 0 & 1 & 0 \\ 0 & 0 & 0 & 0 & 0 & 1 \end{pmatrix} \tag{4.26}$$

and

$$\mathbf{B}_{\text{CHP}} \cdot \mathbf{T}_{0,\text{CHP}} = \begin{pmatrix} -1.2 & 0 & 0 & 0 & 0 & 0 \\ 0.73 & 1.54 & 0 & 0 & 0.55 & 0.48 \end{pmatrix}. \tag{4.27}$$

Before allocation of the non-functional flows of the multi-functional unit process to the duplicated mono-functional unit processes, both the multi-functional and the duplicated mono-functional unit processes require another modification in the matrix $\mathbf{AT}_1$. The multi-functional unit process is represented by the lth column in the matrix $\mathbf{AT}_1$. It should only contain non-functional flows that are allocated to the mono-functional unit processes. For this purpose, the functional flows have to be set to zero in the lth column in the matrix $\mathbf{AT}_1$.

The duplicated mono-functional unit processes are represented by the $(l+1)$th to $(l+\beta)$th column in the matrix $\mathbf{AT}_1$. They should only contain one functional flow each. For this purpose, all economic flows except of one functional flow have to be set to zero in the $(l+1)$th to $(l+\beta)$th duplicated columns of the matrix $\mathbf{AT}_1$.

The modification is realized by an element-wise matrix multiplication of the matrix $\mathbf{AT}_1$ by a *function matrix* $\mathbf{U}$. Herein, the $m \times (n+\beta)$-function matrix $\mathbf{U}$ has the same dimension as the matrix $\mathbf{AT}_1$, i.e.,

$$\mathbf{U} = \begin{pmatrix} u_{11} & \cdots & \cdots & u_{1(n+\beta)} \\ \vdots & \ddots & \ddots & \vdots \\ u_{m1} & \cdots & \cdots & u_{m(n+\beta)} \end{pmatrix}. \tag{4.28}$$

The function matrix $\mathbf{U}$ is constructed from an $(m \times (n+\beta))$-matrix with all elements $u_{ij} = 1$. If the kth to $(k+\beta)$th economic flows in matrix $\mathbf{AT}_1$ are functional flows of the lth multi-functional process, then the function matrix elements u_{kl} to $u_{(k+\beta)l}$ are $u_{kl} = u_{(k+\beta)l} = 0$. This deletes the functional flow in the column of the multi-functional unit process in the matrix $\mathbf{AT}_1$.

In the $l+1$th to $(l+\beta)$th columns of the matrix $\mathbf{U}$, all elements are zero except for the functional flows k t $(k+\beta)$, i.e., $u_{i(l+1)} = u_{i(l+\beta)} = 0$ for all $i \neq k, \ldots, k+\beta$. The matrix elements $u_{k(l+1)}$ to $u_{(k+1)(l+\beta)}$ correspond to the functional flows and are $u_{k(l+1)} = 1$ and $u_{(k+1)(l+\beta)} = 1$. This operation deletes all economic flows except for one functional flow in each duplicated column in the matrix $\mathbf{AT}_1$, thus creating mono-functional processes.

For the illustrating example, the multi-functional unit process is given by the $l = 2$ column of the transformed technology matrix $\mathbf{AT}_1$. The corresponding mono-functional unit processes are represented by the $l = 3$ and $l = 4$ columns in $\mathbf{AT}_1$. The $\beta = 2$ functional flows of the multi-functional unit process are given in the $k = 2$ and $k = 3$ row of $\mathbf{AT}_1$. Following the procedure outlined above, the initial function matrix $\mathbf{U}_{\text{CHP,ini}}$ is modified to $\mathbf{U}_{\text{CHP}}$, i.e.,

$$\begin{aligned} \mathbf{U}_{\text{CHP,ini}} &= \begin{pmatrix} 1 & 1 & \not{1} & \not{1} & 1 & 1 \\ 1 & \not{1} & 1 & \not{1} & 1 & 1 \\ 1 & \not{1} & \not{1} & 1 & 1 & 1 \\ 1 & 1 & \not{1} & \not{1} & 1 & 1 \\ 1 & 1 & \not{1} & \not{1} & 1 & 1 \end{pmatrix} \\ \rightarrow \mathbf{U}_{\text{CHP}} &= \begin{pmatrix} 1 & 1 & 0 & 0 & 1 & 1 \\ 1 & 0 & 1 & 0 & 1 & 1 \\ 1 & 0 & 0 & 1 & 1 & 1 \\ 1 & 1 & 0 & 0 & 1 & 1 \\ 1 & 1 & 0 & 0 & 1 & 1 \end{pmatrix}. \end{aligned} \tag{4.29}$$

An element-wise matrix multiplication is noted with the symbol "$\circ$". Element-wise multiplication of $(\mathbf{AT}_{1,\text{CHP}}$ by $\mathbf{U}_{\text{CHP}}$ yields

$$\mathbf{U}_{\text{CHP}} \circ (\mathbf{AT}_1)_{\text{CHP}} = \begin{pmatrix} 1 & -0.42 & 0 & 0 & -0.15 & -0.13 \\ 0 & 0 & 1 & 0 & 0 & 0 \\ 0 & 0 & 0 & 1.67 & 0 & 0 \\ 0 & 0 & 0 & 0 & 1 & 0 \\ 0 & 0 & 0 & 0 & 0 & 1 \end{pmatrix} \quad (4.30)$$

and concludes the duplication of the multi-functional CHP-process into two mono-functional processes.

4.2.2 Allocating environmental impacts from multi- to mono-functional unit processes

Allocating the non-functional flows of a multi-functional unit process to mono-functional unit processes is realized by a right matrix multiplication of the matrices $\mathbf{U} \circ \mathbf{AT}_1$ and $\mathbf{BT}_0$ by a $(n+\beta) \times m$*-allocation matrix* $\mathbf{C}$:

$$\mathbf{C} = \begin{pmatrix} c_{11} & \cdots & c_{1m} \\ \vdots & \ddots & \vdots \\ \vdots & \ddots & \vdots \\ c_{(n+\beta)1} & \cdots & c_{(n+\beta)m} \end{pmatrix}. \quad (4.31)$$

The allocation matrix is constructed from an initial $(m \times m)$-identity matrix. As in section 4.2.1, the multi-functional process is the lth column of the technology matrix $\mathbf{A}$ and the kth to $(k+\beta)$th rows are its functional flows. Then, a m-dimensional row is inserted to the initial $(m \times m)$-identity matrix. This inserted row is inserted as lth row to the identity matrix. The inserted row contains only zeroes except for the elements in the kth to $k+1$th column: these matrix elements c_{lk} to $c_{l(k+1)}$ represent the allocation factors for the mono-functional processes in the kth and $(k+1)$th column of the matrix $\mathbf{U} \circ \mathbf{AT}_1$.

For the illustrating example, an initial (5×5)-identity matrix is expanded by a row at the $l=2$ position. The second and the third economic flow in the technology matrix A_{CHP} are the functional flows electricity and heat. The corresponding allocation factors are thus $c_{22} = c_{\text{electricity}}$ and $c_{23} = c_{\text{heat}}$. The allocation matrix of the illustrating example is

$$\mathbf{C}_{\text{CHP,ini}} = \begin{pmatrix} 1 & 0 & 0 & 0 & 0 \\ 0 & 1 & 0 & 0 & 0 \\ 0 & 0 & 1 & 0 & 0 \\ 0 & 0 & 0 & 1 & 0 \\ 0 & 0 & 0 & 0 & 1 \end{pmatrix} \rightarrow \mathbf{C}_{\text{CHP}} = \begin{pmatrix} 1 & 0 & 0 & 0 & 0 \\ 0 & c_{\text{electricity}} & c_{\text{heat}} & 0 & 0 \\ 0 & 1 & 0 & 0 & 0 \\ 0 & 0 & 1 & 0 & 0 \\ 0 & 0 & 0 & 1 & 0 \\ 0 & 0 & 0 & 0 & 1 \end{pmatrix}. \quad (4.32)$$

ISO (2006b) requires that the sum of allocated environmental impacts equals the impacts from the multi-functional unit process. This requirement is met if the sum of allocation factors c_{pq} is

$$\sum_{q=1}^{m} c_{pq} = 1 \quad \text{for } p = 1, \ldots, (n+\beta). \quad (4.33)$$

Equation 4.33 can be met using the energy content of the functional flows electricity and heat as allocation criterion. This allocation criterion is used purely for illustration here. The energy content is given by the technology matrix elements a_{22} and a_{32} which represent the economic flows of electricity and eat, respectively. The corresponding allocation factors are

$$c_{\text{electricity}} = \frac{a_{22}}{a_{22}+a_{32}} = 0.37 \quad \text{and} \quad c_{\text{heat}} = \frac{a_{32}}{a_{22}+a_{32}} = 0.63. \quad (4.34)$$

For these allocation factors, a right matrix multiplication of the matrices $(\mathbf{U} \circ (\mathbf{AT}_1))_{\text{CHP}}$ and $(\mathbf{BT}_0))_{\text{CHP}}$ by the allocation matrix $\mathbf{C}_{\text{CHP}}$ yields

$$(\mathbf{U} \circ (\mathbf{AT}_1))_{\text{CHP}} \cdot \mathbf{C}_{\text{CHP}} = \begin{pmatrix} 1 & -0.16 & -0.26 & -0.15 & -0.13 \\ 0 & 1 & 0 & 0 & 0 \\ 0 & 0 & 1.67 & 0 & 0 \\ 0 & 0 & 0 & 1 & 0 \\ 0 & 0 & 0 & 0 & 1 \end{pmatrix} \quad (4.35)$$

and

$$(\mathbf{BT}_0)_{\text{CHP}} \cdot \mathbf{C}_{\text{CHP}} = \begin{pmatrix} -1.2 & 0 & 0 & 0 & 0 \\ 0.73 & 0.57 & 0.97 & 0.55 & 0.48 \end{pmatrix}. \quad (4.36)$$

The matrix $(\mathbf{U} \circ (\mathbf{AT}_1)) \cdot \mathbf{C}$ is a square $(m \times m)$-matrix and allows calculating a scaling vector from

$$\mathbf{s} = ((\mathbf{U} \circ (\mathbf{AT}_1))\,\mathbf{C})^{-1} \cdot \mathbf{f}. \tag{4.37}$$

The corresponding total intervention vector $\mathbf{g}$ and total impact vector $\mathbf{h}$ are obtained from

$$\mathbf{g} = \mathbf{BT}_0\mathbf{C} \cdot ((\mathbf{U} \circ (\mathbf{AT}_1))\,\mathbf{C})^{-1} \cdot \mathbf{f}, \tag{4.38}$$

$$\mathbf{h} = \mathbf{Q} \cdot \underbrace{\mathbf{BT}_0\mathbf{C}}_{\mathbf{B}_{\text{allocated}}} \cdot (\underbrace{(\mathbf{U} \circ (\mathbf{AT}_1))\,\mathbf{C}}_{\mathbf{A}_{\text{allocated}}})^{-1} \cdot \mathbf{f}. \tag{4.39}$$

The subscripted braces in Equation 4.39 illustrate that the new formulas for allocation again have the same structure as those provided by Heijungs and Suh (2002) (cf. Equations 3.7, 3.8, and 3.11). Here, the matrix product $(\mathbf{U} \circ (\mathbf{AT}_1))\,\mathbf{C}$ represents an allocated technology matrix $\mathbf{A}_{\text{allocated}}$; the matrix product $\mathbf{BT}_0\mathbf{C}$ is the corresponding allocated intervention matrix $\mathbf{B}_{\text{allocated}}$.

The number of economic flows remains unaffected by allocation. Therefore, allocation does not modify the original final demand vector $\mathbf{f}$. However, allocation increases the number of unit processes. The dimensions of both allocated technology and intervention matrices $\mathbf{A}_{\text{allocated}}$ and $\mathbf{B}_{\text{allocated}}$ increase compared to the original matrices $\mathbf{A}$ and $\mathbf{B}$. Still, the sparsity of the allocated matrices increases as well. The computational effort for calculating LCA-results with Equations 4.37, 4.38, and 4.39 does not necessarily increase.

4.3 Combined matrix formulation for fixing multi-functionality problems

Industrial process systems often include many multi-functional unit processes and thus as many multi-functionality problems. The different multi-functionality problems can be fixed with different methods, i.e., some multi-functional processes are allocated while others are fixed with avoided burden or system expansion. Therefore, the formulations derived in sections 4.1 and 4.2 are combined into one set of equations allowing different multi-functionality problems to be fixed with different methods.

The combination applies the matrix formulas for allocation to the matrix formulas for system expansion and avoided burden. This combination allows calculating a scaling vector $\mathbf{s}$, LCI-results $\mathbf{g}$, and LCIA-results $\mathbf{h}$ from

$$\begin{aligned}
\mathbf{s} &= \left([\mathbf{U} \circ ((\mathbf{EAV})\,\mathbf{T}_1)]\,\mathbf{C}\right)^{-1} \cdot \mathbf{Ef}, && (4.40)\\
\mathbf{g} &= \mathbf{BVT}_0\mathbf{C} \cdot \mathbf{s} \\
&= \mathbf{BVT}_0\mathbf{C} \cdot \left([\mathbf{U} \circ ((\mathbf{EAV})\,\mathbf{T}_1)]\,\mathbf{C}\right)^{-1} \cdot \mathbf{Ef}, && (4.41)\\
\mathbf{h} &= \mathbf{Q} \cdot \mathbf{g} \\
&= \mathbf{Q} \cdot \underbrace{\mathbf{BVT}_0\mathbf{C}}_{\mathbf{B}_{\text{fixed}}} \cdot \underbrace{\left([\mathbf{U} \circ ((\mathbf{EAV})\,\mathbf{T}_1)]\,\mathbf{C}\right)}_{\mathbf{A}_{\text{fixed}}}^{-1} \cdot \underbrace{\mathbf{Ef}}_{\mathbf{f}_{\text{fixed}}}. && (4.42)
\end{aligned}$$

Here, it is assumed that the change of the functional unit during system expansion (section 4.1.3) has already been implemented and is therefore not displayed in Equations 4.40 to 4.42.

Again, the subscripted braces in Equation 4.39 illustrate that the combined formulas for system expansion, avoided burden and allocation also have the same structure as those provided by Heijungs and Suh (2002) (cf. Equations 3.7, 3.8, and 3.11). Here, the matrix product $(\mathbf{U} \circ ((\mathbf{EAV})\mathbf{T}_1))\,\mathbf{C}$ represents an fixed technology matrix $\mathbf{A}_{\text{fixed}}$. Multi-functionality problems that exist in the original technology matrix $\mathbf{A}$ are fixed in the fixed technology matrix $\mathbf{A}_{\text{fixed}}$. The matrix product $\mathbf{BVT}_0\mathbf{C}$ is the corresponding fixed intervention matrix $\mathbf{B}_{\text{fixed}}$.

If no multi-functionality problem exists, the matrices $\mathbf{E}$, $\mathbf{V}$, $\mathbf{T}_0$, $\mathbf{T}_1$, $\mathbf{U}$, and $\mathbf{C}$ remain identity matrices according to the construction rules described in sections 4.1.1, 4.1.2, 4.2.1, and 4.2.2. In this case, the formulas presented above become identical to the formulas derived by Heijungs and Suh (2002) (section 3.1.3).

For a separate application of system expansion or avoided burden, the transformation, function and allocation matrices $\mathbf{T}_0$, $\mathbf{T}_1$, $\mathbf{U}$, and $\mathbf{C}$ are the initial matrices according to sections 4.2.1 and 4.2.2. The dimensions of the initial matrices are adapted to the dimensions of the equivalence and aggregation matrices $\mathbf{E}$ and $\mathbf{V}$: the transformation matrices $\mathbf{T}_0$ and $\mathbf{T}_1$ remain $(w \times w)$-identity matrices. The function matrix $\mathbf{U}$ is a $(r \times w)$-matrix with all elements $u_{kl} = 1$. Finally, the allocation matrix $\mathbf{C}$ is a $(w \times w)$-identity matrix.

Nomenclature, dimensions, and functionalities of the matrices introduced in this chapter are summarized in Table 4.1.

Table 4.1: Terminology, dimensions and functions of matrices introduced for modeling system expansion, avoided burden, and allocation in matrix-based formulas

symbol	name	dimension	function/content
A	technology matrix	m economic flows × n processes	process data
f	final demand vector	m economic flows	reference flow of FU
B	intervention matrix	v elementary flows × n processes	process data
Q	characterization matrix	u impact flows × v elementary flows	characterization factors
E	equivalence matrix	r economic flows × m economic flows	merging equivalent economic flows
V	aggregation matrix	n processes × w processes	aggregating multiple avoided burden processes
$\mathbf{T}_1$	transformation matrix for technology matrix	w processes × $(w + \beta)$ processes	duplicating columns of multi-functional processes
$\mathbf{T}_0$	transformation matrix for intervention matrix	w processes × $(w + \beta)$ processes	duplicating columns of multi-functional processes
U	function matrix	r economic flows × $(w + \beta)$ processes	creating mono-functional processes
C	allocation matrix	$(w + \beta)$ processes × r processes	allocating non-functional flows to mono-functional processes

4.4 Discussion and outlook

Equations 4.40, 4.41, and 4.42 are the final results of this chapter. These equations form an expanded matrix formulation for calculating LCA-results. In contrast to the underlying equations from Heijungs and Suh (2002), the expanded formulas include explicit functional relationships for fixing multi-functionality problems with three methods: System expansion, avoided burden, and allocation are all implemented in the expanded formulation.

The expanded formulation could be directly implemented in a LCA software tool that employs the matrix formulation, e.g., CMLCA (Heijungs, 2013a) or Simapro (PRé Consultants, 2013). An implementation in existing LCA software tools would allow to further test the applicability of the proposed formulation for real-life LCAs. It could be particularly interesting to investigate process systems with several cases of multi-functionality. For this case, the presented operations to construct the matrices $\mathbf{E}$, $\mathbf{V}$, $\mathbf{T}_1$, $\mathbf{T}_0$, $\mathbf{U}$, and $\mathbf{C}$ have to be executed several times. In future work, it should be tested if real-world LCAs can be calculated with similar computational effort when the expanded formulas are applied.

The formulas require inputs from LCA-practitioners. The inputs are similar to those required by existing software tools when a multi-functionality problem is fixed:

For system expansion and avoided burden, the LCA-practitioner needs to decide which added or avoided processes should be used. This choice has to be taken in every LCA software tool. The aggregation of multiple added or avoided processes into one unit process is not a common operation in existing software tools. With existing software tools, the procedure typically depends on whether a marginal or a mix process is used as added or avoided process.

For marginal processes, LCA-practitioners commonly calculate separate LCA-results for each marginal candidate process for an added or avoided process. This procedure can be maintained by setting the aggregation factor for the selected candidate process to one and all other aggregation factors to zero.

Mix processes are directly provided by LCI-databases such as ecoinvent (Frischknecht et al., 2005) or GaBi (PE International, 2013). Those unit processes are based on a constant mix composition, e.g., the electricity mix of a country for a specified time period. The matrix formulation proposed in this work requires that the LCA-practitioner specifies the mix composition in the aggregation factors. However, providers of LCA-databases could also adapt their datasets and provide a mix process as a set of unit processes along with aggregation factors defining the mix composition.

The required input for allocation is identical to the input for allocation in existing software tools: The LCA-practitioner only needs to specify the allocation criterion, i.e., the allocation factors. Here, it is assumed that the functional flows of each unit process are known.[3]

[3]In common LCI-databases such as ecoinvent (Frischknecht et al., 2005), the functional flows of each unit process are known and specified.

In section 4.2 and Equations 4.40 to 4.42, one allocation criterion is applied to all non-functional flows that are allocated. In general, this is, however, a simplification. In practice, multiple allocation criteria can be used for allocation of one multi-functional unit process: E.g., energy-related non-functional flows are allocated according to the energy-content of the functional flows while carbon-related non-functional flows are allocated using the carbon-content of the functional flows as criterion. In future work, the formulas derived in section 4.2 should be expanded to include multiple allocation criteria for allocation of one multi-functional unit process. This could be done by transforming a multi-functional unit process with β functional flows into $x \cdot \beta$ mono-functional unit processes. Here, x refers to the number of different allocation criteria that are applied.

The aggregation factors v_{xy} and the allocation factors c_{xy} are the key parameters that allow analytical analysis of uncertainties due to fixing multi-functionality problems. The expanded matrix formulation is used to derive a procedure for fixing multi-functionality problems in comparative LCA of multi-product processes in chapter 5. In chapter 6, a holistic uncertainty analysis method based on analytical error propagation is derived from Equations 4.40 to 4.42.

Chapter 5

A procedure for fixing multi-functionality problems in comparative LCA

Comparative LCA of multi-product processes causes a multi-functionality problem (section 3.2.4). The problem occurs in particular for comparisons of multi-product processes with non-common products such as the chlor-alkali electrolysis processes discussed in chapter 2. In this chapter, a procedure for fixing multi-functionality problems in comparative LCA of multi-product processes with non-common products is derived. An earlier version of the procedure is published by the author of this work (Jung et al., 2013a).

First, the mathematical structure of multi-functionality in comparative LCA is discussed in section 5.1. For this purpose, a matrix-based formulation for comparative LCA is introduced in section 5.1.1. The mathematical definition of multi-functionality problems in comparative LCA is subsequently described in section 5.1.2.

In section 5.2, a procedure for fixing multi-functionality problems in comparative LCA is derived. The procedure distinguishes main products and by-products (section 5.2.1). It consists of a two-part workflow providing recommendations for fixing multi-functionality problems due to non-common main products (section 5.2.2) and non-common by-products (section 5.2.3). The chapter concludes with a discussion of the procedure in section 5.3.

5.1 Multi-functionality in comparative LCA

5.1.1 A matrix formulation of comparative LCA

Comparative LCA analyzes alternative products or alternative processes (section 3.1.2). The general focus of this work is on alternative processes producing the same products. Still, the following explanations apply to alternative products as well.

A matrix based formulation for comparative LCA has already been presented (Heijungs and Suh (2002) on p. 84). In this section, that formulation by Heijungs and Suh (2002) is rewritten with the expanded formulas derived in the previous chapter 4 of this work.

In comparative LCA, the analyzed alternative unit processes are included in one technology matrix **A** and one corresponding intervention matrix **B**. The functional flows of those alternative unit processes are economic flows satisfying the same functional unit. Those economic flows are represented by different rows in the technology matrix **A**. Therefore, each studied alternative unit process corresponds to a different final demand vector (Heijungs and Suh (2002) on p. 84).

In comparative LCA, a total number of Λ alternative final demand vectors $\mathbf{f}_\lambda$ are summarized in a $(m \times \Lambda)$-*final demand matrix* **F**, i.e.,

$$\mathbf{F} = \left(\mathbf{f}_1 \,\middle|\, \cdots \,\middle|\, \mathbf{f}_\Lambda \right). \tag{5.1}$$

Comparative LCA-results are collected in a $(n \times \Lambda)$-*scaling matrix* **S**, an $(v \times \Lambda)$-*total intervention matrix* **G**, and an $(u \times \Lambda)$-*total impact matrix* **H**. Those matrices are calculated using Equations 4.40, 4.41, and 4.42, i.e.,

$$\begin{aligned}
\mathbf{S} &= \left(\mathbf{s}_1 \,\middle|\, \cdots \,\middle|\, \mathbf{s}_\Lambda \right) \\
&= \left(\left[\mathbf{U} \circ \left(\left(\mathbf{EAV} \right) \mathbf{T}_1 \right) \right] \mathbf{C} \right)^{-1} \cdot \mathbf{EF}, && (5.2) \\
\mathbf{G} &= \left(\mathbf{g}_1 \,\middle|\, \cdots \,\middle|\, \mathbf{g}_\Lambda \right) \\
&= \mathbf{BVT}_0\mathbf{C} \cdot \left(\left[\mathbf{U} \circ \left(\left(\mathbf{EAV} \right) \mathbf{T}_1 \right) \right] \mathbf{C} \right)^{-1} \cdot \mathbf{EF}, && (5.3) \\
\mathbf{H} &= \left(\mathbf{h}_1 \,\middle|\, \cdots \,\middle|\, \mathbf{h}_\Lambda \right) \\
&= \mathbf{Q} \cdot \mathbf{BVT}_0\mathbf{C} \cdot \left(\left[\mathbf{U} \circ \left(\left(\mathbf{EAV} \right) \mathbf{T}_1 \right) \right] \mathbf{C} \right)^{-1} \cdot \mathbf{EF}. && (5.4)
\end{aligned}$$

Each **S**, **G**, and **H** includes a total number of Λ corresponding vectors $\mathbf{s}_\lambda$, $\mathbf{g}_\lambda$, and $\mathbf{h}_\lambda$. Those vectors are the LCA-results of each alternative unit process.

5.1.2 Multi-functionality problems in comparative LCA

In this section, multi-functionality problems in comparative LCA are defined. For this purpose, the definition of multi-functionality problems from Frischknecht (1998) is applied to the formulas for comparative LCA presented in the previous section.

For comparative LCA, a $(m \times \Lambda)$-*discrepancy matrix* $\mathbf{D}$ consists of the corresponding discrepancy vectors $\mathbf{d}$ (cf. Equation 3.16), i.e.,

$$\mathbf{D} = \left(\mathbf{d}_1 \,\middle|\, \cdots \,\middle|\, \mathbf{d}_\Lambda \right) = \tilde{\mathbf{F}} - \mathbf{F} = \mathbf{A} \cdot \mathbf{A}^{+}\mathbf{F} - \mathbf{F}. \tag{5.5}$$

Here, the $(m \times \Lambda)$-*final supply matrix* $\tilde{\mathbf{F}}$ is composed of the corresponding final supply vectors $\tilde{\mathbf{f}}$, i.e.,

$$\tilde{\mathbf{F}} = \left(\tilde{\mathbf{f}}_1 \,\middle|\, \cdots \,\middle|\, \tilde{\mathbf{f}}_\Lambda \right). \tag{5.6}$$

Following Equation 3.18, a multi-functionality problems for comparative LCA exists if

$$\mathbf{D} \neq \mathbf{0}^{m \times \Lambda}. \tag{5.7}$$

The $(m \times \Lambda)$-matrix $\mathbf{0}$ is called zero-matrix because all matrix elements are $0_{i\lambda} = 0$.

If a multi-functionality problem exists, the discrepancy matrix contains information on the economic flows that cause the multi-functionality problem. The elements $d_{i\lambda}$ represent the difference between final supply $\tilde{f}_{i\lambda}$ and final demand $f_{i\lambda}$ of each economic flow, i.e.,

$$d_{i\lambda} = \tilde{f}_{i\lambda} - f_{i\lambda}. \tag{5.8}$$

A surplus of an economic flow is identified by $d_{i\lambda} > 0$ (Marvuglia et al., 2010). In contrast, the final demand $f_{i\lambda}$ exceeds the final supply $\tilde{f}_{i\lambda}$ of an economic flow if $d_{i\lambda} < 0$.

Identifying surplus economic flows is important because these economic flows are involved in all methods for fixing the multi-functionality problem. For system expansion, the functional unit has to be expanded to include the economic flows with a surplus. For avoided burden, an avoided process is required producing an economic flow equivalent to the surplus flow. For allocation, environmental impacts have to be allocated between functional flows and an economic flows with a surplus.

In this chapter, comparative LCA is illustrated using a process system shown in the process flow diagram in Figure 5.1. The process system includes three unit processes: mining (process 1), electricity generation (process 2), and CHP (process 3). The electricity generation process and the CHP-process are two alternative processes for the production of electricity.

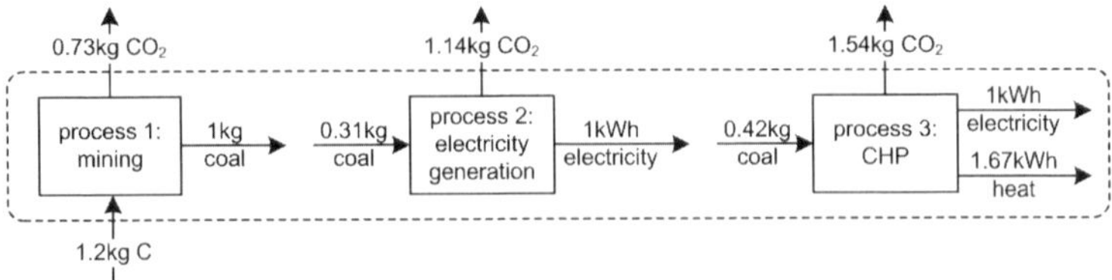

Figure 5.1: Process flow diagram of a process system with three unit processes: mining, electricity generation, and CHP. Electricity generation and CPP are alternative processes for the production of electricity.

This process system's technology matrix $\mathbf{A}_{\text{el}}$ is

$$\mathbf{A}_{\text{el}} = \begin{pmatrix} 1 & -0.31 & -0.42 \\ 0 & 1 & 0 \\ 0 & 0 & 1 \\ 0 & 0 & 1.67 \end{pmatrix}. \tag{5.9}$$

The index of this technology matrix refers to the common functional flow of both alternative processes, i.e., *electricity* (el).

Comparative LCA-studies focusing on alternative products have a precise functional unit, e.g., "producing 1000 kWh electricity" for the product electricity. But comparative LCA-studies focusing on alternative processes allow for different functional units depending on the scope of the study. E.g., the electricity generation process and the CHP-process can be compared based on two functional units: "producing 1000 kWh electricity" and "producing 1000 kWh electricity and 1670 kWh heat". The corresponding final demand matrices $\mathbf{F}_{\text{el}}$ and $\mathbf{F}_{\text{el+he}}$ are

$$\mathbf{F}_{\text{el}} = \begin{pmatrix} 0 & 0 \\ 1000 & 0 \\ 0 & 1000 \\ 0 & 0 \end{pmatrix} \tag{5.10}$$

and

$$\mathbf{F}_{\text{el+he}} = \begin{pmatrix} 0 & 0 \\ 1000 & 0 \\ 0 & 1000 \\ 1670 & 1670 \end{pmatrix}. \quad (5.11)$$

Here, the index of the final demand matrix $\mathbf{F}_{\text{el+he}}$ refers to both functional flows of the CHP-process, i.e., electricity and *heat* (he).

Both final demand matrices cause a multi-functionality problem. For the final demand matrix $\mathbf{F}_{\text{el}}$, the corresponding discrepancy matrix $\mathbf{D}_{\text{el}}$ is

$$\mathbf{D}_{\text{el}} = \begin{pmatrix} 0 & 0 \\ 0 & 0 \\ 0 & -736 \\ 0 & 441 \end{pmatrix}. \quad (5.12)$$

In this discrepancy matrix $\mathbf{D}_{\text{el}}$, $d_{42} > 0$ indicates a surplus of heat from the CHP process (process 3 in Figure 5.1.

For the final demand matrix $\mathbf{F}_{\text{el+he}}$, the corresponding discrepancy matrix $\mathbf{D}_{\text{el+he}}$ is

$$\mathbf{D}_{\text{el+he}} = \begin{pmatrix} 0 & 0 \\ 0 & 0 \\ 0.736 & 0 \\ -0.441 & 0 \end{pmatrix}. \quad (5.13)$$

In this discrepancy matrix $\mathbf{D}_{\text{el+he}}$, $d_{31} > 0$ indicates a surplus of electricity from the CHP-process (process 3 in Figure 5.1. Moreover comparison of the processes 2 and 3 in Figure 5.1 is not achieved with the final demand matrix $\mathbf{F}_{\text{el+he}}$, because process 3 is supplying heat for both compared alternatives. Here, the intuitively chosen final demand matrix $\mathbf{F}_{\text{el+he}}$ causes a multi-functionality problem and does not correspond with the scope of comparing process 2 and 3 in Figure 5.1.

In these two examples, the multi-functionality problems exists because a multi-functional unit process (process 3 in Figure 5.1) is compared to a mono-functional process (process 2 in Figure 5.1). The two compared processes have one common functional flow, i.e., electricity. But they also have a non-common functional flow, i.e., heat. Non-common functional flows are the reason for multi-functionality problems during comparative LCA of alternative unit processes. Non-common functional flows also exists when comparing the membrane and the ODC-processes for chlor-alkali electrolysis (cf. chapter 2). Both membrane and ODC-processes have common products, i.e., chlorine and caustic soda while hydrogen is a non-common product.

The final demand matrix $\mathbf{F}_{\mathrm{el+he}}$ can be regarded as system expansion because it includes the additional though non-common function of heat production. The example with the final demand matrix $\mathbf{F}_{\mathrm{el+he}}$ illustrates that multi-functionality problems due to non-common functional flows differ from multi-functionality problems in non-comparative LCA: A multi-functionality problem in non-comparative LCA can always be fixed with system expansion (section 3.2.3.1).[1] For comparative LCA, system expansion using the final demand matrix $\mathbf{F}_{\mathrm{el+he}}$ fails, because $\mathbf{D}_{\mathrm{el+he}} \neq \mathbf{0}$. Thus, system expansion only fixes a multi-functionality problem in comparative LCA if added unit processes are used.

The existing recommendations for fixing multi-functionality problems advise to avoid allocation by system expansion (ISO (2006b) on p. 29, European Commission (2010b) on pp. 72-81). But system expansion can fail to fix multi-functionality problems in comparative LCA as discussed above. Therefore, a procedure for fixing multi-functionality problems in comparative LCA is derived in the following section 5.2.

5.2 Fixing multi-functionality problems in comparative LCA

The procedure for fixing multi-functionality problems begins with a distinction of product flows in main products and by-products. The distinction is used as criterion to decide if non-common functional flows should be included in the functional unit (section 5.2.1). In the subsequent sections 5.2.2 and 5.2.3, recommendations are given how to fix multi-functionality problems in comparative LCA conform to ISO (2006a).

5.2.1 Flow distinction in comparative LCA of alternative multi-product processes with non-common products

In the previous section, it is explained that multi-functionality problems occur in comparative LCA whenever the compared alternative processes have non-common functional flows. *Common functional flows* are flows that have equal quantities and

[1] In practice, system expansion can still be infeasible because the change of the functional unit may contradict with the scope of a LCA-study.

properties (e.g., thermodynamic state of a material flow, temperature of a heat flow). Consequently, non-common refer not only to the existence of functional flows but also to differences in their quantity and their properties.

Functional flows of multi-product processes are product output flows, cf. section 3.2.1. In this work, products are distinguished in *main products* and *by-products*.[2] Main products are the economic reason for process operation, i.e., a unit process is operated because of the economic value of the main products. In contrast, by-products still have a positive economic value but a unit process is not installed and operated because of a by-product's economic value. The distinction between main products and by-products can vary depending on economic and technical framework conditions. Examples of economic framework conditions are market prices, subsidies, or tax regulations. An example for technical framework conditions is whether infrastructure for product transportation and use is available or not.

The distinction between main products and by-products allows for systematic functional unit definition in comparative LCA of alternative processes: The functional unit should include all - common and non-common - main products of the alternative processes. Including all main products in the functional unit ensures that the environmental impacts of alternative processes are compared based on their economic reason of operation. This suggestion seems reasonable because the economic reason of operation determines whether a process is actually operated or not and thus causes environmental impacts.

The distinction between main products and by-products does not fix multi-functionality problems due to non-common products yet. But multi-functionality problems can now be related to either non-common main products or non-common by-products. It is argued in the following section 5.2.2 that a multi-functionality problem due to non-common main products should be fixed with system expansion. In section 5.2.3, recommendations are given for fixing multi-functionality problems due to non-common by-products.

5.2.2 Fixing multi-functionality problems due to non-common main-products

In this section, the first part of the procedure for fixing multi-functionality problems is introduced. The first part refers to multi-functionality problems due to non-

[2]The guide from the EU-commission uses the broader terms determining co-function and depending co-function instead of main product and by-product (European Commission, 2010b).

common main products. The workflow comprises a systematic method identifying multi-functionality problems due to non-common main products and suggests to fix those problems with system expansion. The first part of the procedure is illustrated in a workflow (Figure 5.2) and explained in the following paragraphs.

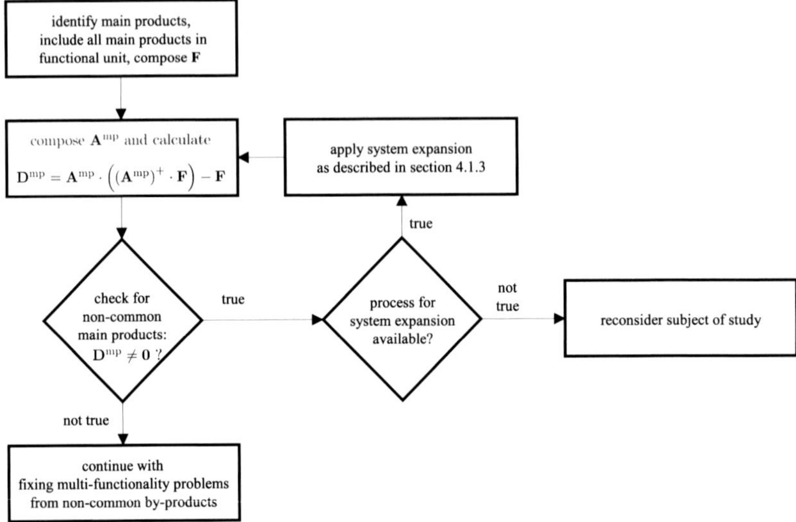

Figure 5.2: Procedure for fixing multi-functionality problems in comparative LCA: first part of a workflow for identifying and fixing multi-functionality problems due to non-common main products.

In the first step, the main products are determined and included in the functional unit (section 5.2.1). A final demand matrix $\mathbf{F}$ is composed of final demand vectors $\mathbf{f}_\lambda$ for each alternative process. Each final demand vector $\mathbf{f}_\lambda$ includes all main product flows of the corresponding alternative process λ.

In the following steps, it is checked if a multi-functionality problem due to non-common main products exists. For this purpose, non-common main products have to identified. Non-common main products could be identified manually by comparing the existence, the quantities, and the qualities of the functional flows in those final demand vectors $\mathbf{f}_\lambda$. A more systematic method uses a called main product discrepancy matrix $\mathbf{D}^{mp}$ that is calculated from the final demand matrix $\mathbf{F}$ and a so-called main product technology matrix $\mathbf{A}^{mp}$, i.e.,

$$\mathbf{D}^{\mathrm{mp}} = \mathbf{A}^{\mathrm{mp}} \cdot \left((\mathbf{A}^{\mathrm{mp}})^{+} \cdot \mathbf{F} \right) - \mathbf{F} \tag{5.14}$$

Here, main product (mp) is indicated by the superscript of the matrices $\mathbf{D}^{\mathrm{mp}}$ and $\mathbf{A}^{\mathrm{mp}}$. The main product technology matrix $\mathbf{A}^{\mathrm{mp}}$ is obtained from the technology matrix $\mathbf{A}$ by deleting all economic flows that are not main products of the alternative unit processes.

Based on the general definition for a multi-functionality problem in comparative LCA (Equation 5.7), a multi-functionality problem due to non-common main products exists if

$$\mathbf{D}^{\mathrm{mp}} \neq \mathbf{0}. \tag{5.15}$$

If Equation 5.15 is not true, no multi-functionality problem due to non-common main products exists and the first part of the workflow is already complete.

However, if Equation 5.15 is true, a multi-functionality problem due to non-common main products needs to be fixed. For this purpose, it is suggested to use system expansion to be conform with the ISO-standard.

System expansion is also suggested because of the functional unit criterion recommending all main products to be included in the functional unit (section 5.2.1). Allocation is less favorable because it would create mono-functional unit processes for the common and non-common main products. The mono-functional unit process for the non-common main product would then be used for all compared alternative unit processes and thereby foiling the intended comparison.

Still, system expansion requires an added process(es) producing the non-common main product(s). The selection of an added process can follow the recommendations in European Commission (2010b) which uses marginal or mix processes (European Commission (2010b) on p. XX). If no added process is found for the non-common main product, system expansion fails and the subject of a study has to be reconsidered.

If an added process exists, system expansion can be applied following section 4.1.3. If added processes are multi-functional themselves, other than the non-common main product flow should be considered by-products because they are not involved in the reason of operation of the compared unit processes. The main product discrepancy matrix $\mathbf{D}^{\mathrm{mp}}$ is thus always $\mathbf{D}^{\mathrm{mp}} = \mathbf{0}$ after system expansion. The first part of the workflow is complete and the procedure continues with by-products in section 5.2.3.

For illustration, the first part of the workflow is applied to the example sketched in section 5.1.1. The functional unit and its corresponding final demand vectors depend on the distinction between main and by-products. In this work, it is assumed that electricity is always a main product of the processes 2 and 3 in Figure 5.1. Therefore, two functional units are possible depending on the classification of heat.

If electricity is the only main product, the functional unit should be, e.g., "producing 1000 kWh electricity" and the corresponding final demand matrix is given in Equation 5.10. Moreover, the main product technology matrix $\mathbf{A}^{\text{mp}}_{\text{el}}$ is

$$\mathbf{A}^{\text{mp}}_{\text{el}} = \begin{pmatrix} 0 & 0 & 0 \\ 0 & 1 & 0 \\ 0 & 0 & 1 \\ 0 & 0 & 0 \end{pmatrix}. \tag{5.16}$$

A multi-functionality problem due to non-common main products does not exist because $\mathbf{D}^{\text{mp}}_{\text{el}} = \mathbf{0}$.

If electricity and heat are both main products of process 3 in Figure 5.1, the functional unit should be, e.g., "producing 1000 kWh electricity and 1670 kWh heat". The corresponding final demand matrix is given in Equation 5.11 and the main product technology matrix is

$$\mathbf{A}^{\text{mp}}_{\text{el+he}} = \begin{pmatrix} 0 & 0 & 0 \\ 0 & 1 & 0 \\ 0 & 0 & 1 \\ 0 & 0 & 1.67 \end{pmatrix}. \tag{5.17}$$

The main product discrepancy matrix $\mathbf{D}^{\text{mp}}_{\text{el+he}}$ becomes

$$\mathbf{D}^{\text{mp}}_{\text{el+he}} = \begin{pmatrix} 0 & 0 \\ 0 & 0 \\ 0.736 & 0 \\ -0.441 & 0 \end{pmatrix}. \tag{5.18}$$

Here, a multi-functionality problem due to the non-common main product heat exists. This multi-functionality problem cannot be fixed using the final demand matrix Equation 5.11 and the processes shown in Figure 5.1. System expansion requires an added process for heat generation.

A process system with two added processes for the production of heat is shown in the process flow diagram in Figure 5.3. In addition to the three unit processes described in Figure 5.1, the expanded process system includes and old boiler (process

4) and a new boiler (process 5) for the production of heat. The technology matrix $\mathbf{A}_{\text{el,exp}}$ of the process system shown in Figure 5.3 is

$$\mathbf{A}_{\text{el,exp}} = \begin{pmatrix} 1 & -0.31 & -0.42 & -0.15 & -0.13 \\ 0 & 1 & 0 & 0 & 0 \\ 0 & 0 & 1 & 0 & 0 \\ 0 & 0 & 1.67 & 0 & 0 \\ 0 & 0 & 0 & 1 & 0 \\ 0 & 0 & 0 & 0 & 1 \end{pmatrix}. \tag{5.19}$$

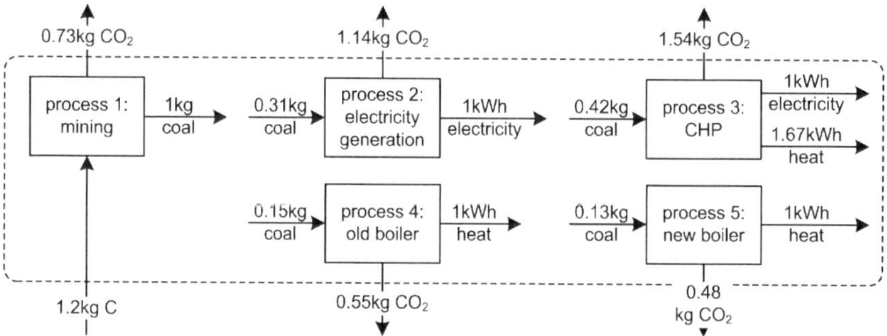

Figure 5.3: Process flow diagram of process system with 5 unit processes: mining, electricity generation, CHP, old boiler, and new boiler. The electricity generation process and the CHP-process are two alternative processes for the production of electricity. The old boiler process and the new boiler process are alternative processes for the production of heat.

The expanded process system's main product technology matrix $\mathbf{A}^{\text{mp}}_{\text{el+he,exp}}$ is

$$\mathbf{A}^{\text{mp}}_{\text{el+he,exp}} = \begin{pmatrix} 0 & 0 & 0 & 0 & 0 \\ 0 & 1 & 0 & 0 & 0 \\ 0 & 0 & 1 & 0 & 0 \\ 0 & 0 & 1.67 & 0 & 0 \\ 0 & 0 & 0 & 1 & 0 \\ 0 & 0 & 0 & 0 & 1 \end{pmatrix}, \tag{5.20}$$

including two heat flows as additional main products. For illustration, it is assumed that only process 5 is used as added process. The expanded final demand matrix $\mathbf{F}_{\text{el+he,exp}}$ becomes

$$\mathbf{F}_{\text{el+he,exp}} = \begin{pmatrix} 0 & 0 \\ 1000 & 0 \\ 0 & 1000 \\ 0 & 1670 \\ 0 & 0 \\ 1670 & 0 \end{pmatrix}. \tag{5.21}$$

Recalculating a main product discrepancy matrix $\mathbf{D}^{\text{mp}}_{\text{el+he,exp}}$ from $\mathbf{A}^{\text{mp}}_{\text{el+he,exp}}$ and $\mathbf{F}_{\text{el+he,exp}}$ yields

$$\mathbf{D}^{\text{mp}}_{\text{el+he,exp}} = \mathbf{0} \tag{5.22}$$

and confirms that the multi-functionality problem from non-common main products is fixed.

If both process 4 and 5 in Figure 5.3 are considered suitable added processes, the procedures for merging and aggregation outlined in sections 4.1.1 and 4.1.2 can be applied.

The application of the workflow to the example demonstrates the systematic identification of multi-functionality problems due to non-common main products in comparative LCA. Moreover, it shows that system expansion requires added processes to fix such multi-functionality problems. It also illustrates that LCA-practitioners face two decisions when applying the workflow: Identifying the main product(s) and selecting the added process. But there is no choice between the methods for fixing multi-functionality problems.

In the next section, the procedure continues with the second part of the workflow.

5.2.3 Fixing multi-functionality problems due to common and non-common by-products

The first part of the workflow ensures that multi-functionality problems due to non-common main products are fixed. In the this section, the second part of the workflow is introduced. The second part refers to multi-functionality problems due to non-common and common by-products. It is suggested to fix those multi-functionality problems due to non-common by-products preferably with avoided burden. Allocation is recommended for those multi-functionality problems due to common by-products and in case that an avoided process does not exist for a non-common by-product.

The second part of the workflow is shown in Figure 5.4 and is further explained in the following paragraphs.

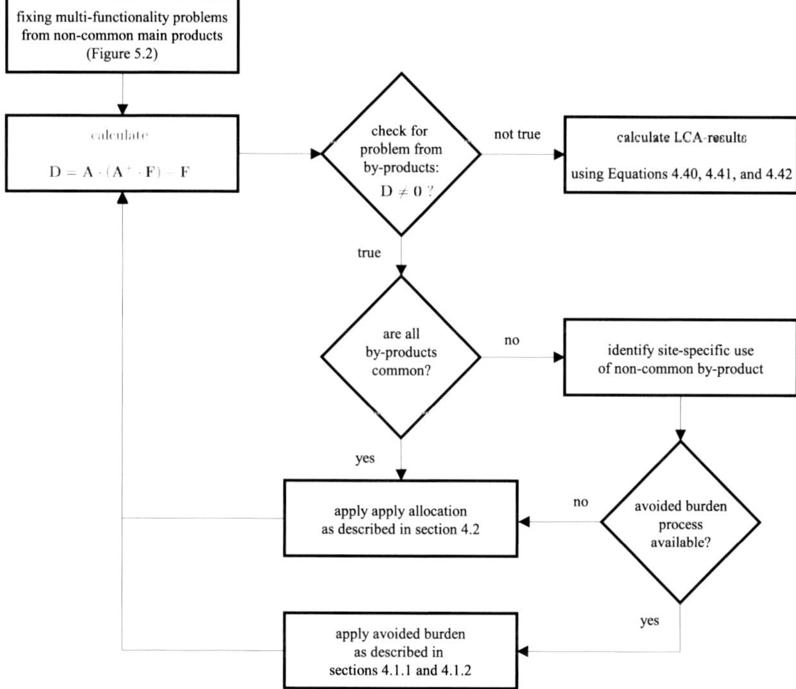

Figure 5.4: Procedure for fixing multi-functionality problems in comparative LCA: second part of workflow for identifying and fixing multi-functionality problems due to non-common by-products.

The second part of the workflow continues with first calculates the discrepancy matrix $\mathbf{D}$. If $\mathbf{D} = \mathbf{0}$, no multi-functionality problem exists and LCA-results can be calculated from Equations 4.40 to 4.42. But for $\mathbf{D} \neq \mathbf{0}$, a multi-functionality problem exists (cf. Equation 5.7). This multi-functionality problem must be related to non-common by-products, because multi-functionality problems due to non-common main products are already fixed in the first part of the workflow.

If Equation 5.7 is true, a positive discrepancy matrix elements $d_{i\lambda} > 0$ identifies a surplus of the ith economic flow (cf. Equation 5.8). That ith economic flow is a by-product causing a multi-functionality problem. In the next step of the workflow, it has to be checked if the by-product causing a multi-functionality is common or non-common.

In contrast to main products, both non-common and also common by-products can cause multi-functionality problems. A common by-product exists if the surpluses $d_{i\lambda} > 0$ are equal for all Λ alternative processes. In contrast, $d_{i\lambda} > 0$ and $d_{i\lambda} \neq d_{i(\lambda+1)}$ indicate a non-common by-product.

If by-products causing a multi-functionality problem are common to all alternative unit processes, it is suggested to apply allocation for fixing that multi-functionality problem. When by-products are common to all compared alternative unit processes, the selected method for fixing a multi-functionality problem influences the absolute LCA-results of each alternative unit process but does not influence the relative result of the comparison. Multi-functionality problems due to common by-products are thus less important for comparative LCA. Therefore, it is suggested to fix multi-functionality problems due to common by-products with allocation. Allocation is quickly applied and avoids additional data collection compared to system expansion and avoided burden. For common by-products of the compared alternative processes, economic allocation is favorable because economic information is already used for the distinction of main products and by-products (section 5.2.1).

It is suggested to fix multi-functionality problems due to non-common by-products with avoided burden conforming to ISO (2006b) and European Commission (2010b). Here, avoided burden is preferred over system expansion because system expansion could change the functional unit and thereby violate that only main products should be included in the functional unit (section 5.2.1).

An avoided process should reflect the non-common by-products site-specific use (Weidema, 2000; European Commission, 2010b). A by-product's use can vary depending on the site-specific conditions. Therefore, the workflow requires identifying the non-common by-products site-specific use before applying avoided burden. For a given site-specific use of a non-common by-product, an avoided process should be selected following European Commission (2010b), i.e., choosing either a marginal or a mix process.

If no avoided process exists, the workflow suggests to fix a multi-functionality problem with allocation. Here, economic allocation is favorable because economic information is already used for the distinction of main products and by-products (section 5.2.1).

After avoided burden or allocation is applied, the workflow requires recalculation of the discrepancy matrix $\mathbf{D}$. If $\mathbf{D} = \mathbf{0}$, the multi-functionality problem due to non-common by-products is fixed and the workflow concludes with the calculation of the LCA-results from Equations 4.40 to 4.42.

If $\mathbf{D} \neq \mathbf{0}$ after applying avoided burden, a multi-functionality problem due to non-common by-products still exists because the avoided process is a multi-product process itself and has introduced a new non-common by-product. Using multi-product processes for avoided burden may potentially cause an endless series of loops. An endless loop can be aborted by applying allocation. It is suggested to apply allocation once the economic or technical relevance of newly introduced by-products is below a certain value.

In this work, a pragmatic approach for checking the relevance of newly introduced by-products is suggested. The approach applies "50:50"-allocation criterion for the multi-functional avoided process[3]. The significance of a newly introduced by-product is low if LCA-results obtained from "50:50"-allocation differ only slightly (e.g., max 1% difference in relevant impact categories) from LCA-results calculated using an allocation factor of zero for the newly introduced non-common by-product.

After an endless loop of avoided burden is aborted with allocation, recalculation of the discrepancy matrix $\mathbf{D}$ always yields $\mathbf{D} = \mathbf{0}$. The second part of the workflow and thereby the entire procedure thus concludes with calculation of LCA-results.

The second part of the workflow is also applied to the illustrating example (Figure 5.3). As in the illustration of the first part of the workflow (section 5.2.2, two final demand matrices have to be distinguished.

If electricity and heat are both main products (final demand matrix $\mathbf{F}_{\text{el+he}}$ in Equation 5.21), there are simply no by-products. The discrepancy matrix is $\mathbf{D}_{\text{el+he}} = \mathbf{0}$ and LCA-results can be calculated from Equations 4.40 to 4.42.

In contrast, a non-common by-product exists if electricity is the only main product (final demand matrix $\mathbf{F}_{\text{el}}$ in Equation 5.10). The discrepancy matrix $\mathbf{D}_{\text{el}}$ is given in Equation 5.12. Here, $d_{42} > 0$ indicates a surplus of the by-product heat which causes a multi-functionality problem.

The by-product heat is non-common, thus avoided burden should be applied to fix this multi-functionality problem. For illustration, it is assumed that the site-specific use of the by-product heat is heat supply to an industrial process. Candidates for avoided processes could be the old and new boilers (process 4 and 5 in Figure 5.3). Applying avoided burden thus requires merging the 3rd row of the technology matrix

[3] "50:50"-allocation is a common term for using an equal allocation factor (Klöpffer and Grahl, 2009)

$\mathbf{A}_{\mathrm{el,exp}}$ (Equation 5.19) with the 4th or 5th row following chapter 4. Instead of selecting exactly one avoided burden process, the candidate processes 4 and 5 in Figure 5.3 can also be aggregated as described in chapter 4.

Nevertheless, both processes 4 and 5 are mono-functional. Therefore, a recalculation of the discrepancy after applying avoided burden yields a discrepancy matrix $\mathbf{D}_{\mathrm{el,exp}} = \mathbf{0}$. The workflow concludes with the calculation of LCA-results using Equations 4.40 to 4.42.

The complete procedure for fixing multi-functionality problems in comparative LCA consisting of the two-part workflow is shown in Figure 5.5. The advantages and potential problems of the procedure are briefly discussed in the following section.

5.3 Discussion and outlook

The presented procedure for fixing multi-functionality problems in comparative LCA consists of a two part workflow (Figure 5.5). The workflow systematically uses the discrepancy matrix (Equation 5.5) to identify multi-functionality problems due to non-common products. Likewise, identification of those multi-functionality problems corresponds to "partial equivalence" in comparative LCA (European Commission (2010b) on p. 67). So far, a precise criterion for "partial equivalence" has been missing.

The workflow suggests how to fix multi-functionality problems due to non-common products, i.e., how to "render comparability" (European Commission (2010b) on p. 67). The workflow conforms to ISO (2006b) and European Commission (2010b), it prefers system expansion and avoided burden over allocation. In contrast to European Commission (2010b), the workflow is tailored to comparative LCA-studies of multi-product processes and clearly distinguishes between system expansion and avoided burden: System expansion should be applied for multi-functionality problems due to non-common main products, avoided burden for problems due to non-common by-products. The presented workflow thus complements ISO (2006b) and European Commission (2010b) for comparative LCA of multi-product processes.

The recommendation how to fix multi-functionality problems in comparative LCA is based on the distinction between main and by-products. This distinction is also used for a suggestion how to choose the functional unit in comparative LCA of multi-product processes: The functional unit should include all main products of the compared alternative unit processes.

A similar procedure as the workflow shown in Figure 5.5 could be derived for other types of multi-functional unit processes (section 3.2.1). An analog distinction between *main wastes* and *by-wastes* seems necessary. However, the elaboration of such a workflow is beyond the scope of this work.

Application of the workflow leaves the following decisions for the LCA-practitioner: distinguishing between main and by-products; selecting added or avoided processes; selecting allocation criterion.

The distinction between main and by-products requires knowledge of the economic settings. Indeed, the distinction may vary from site to site depending on those economic settings. In chapter 7, the workflow is applied to comparative LCA of the chlor-alkali electrolysis introduced in chapter 2. There, it is shown that varying economic settings require evaluation of different scenarios.

Selecting added or avoided processes and allocation factors may follow European Commission (2010b). However, all these selections may introduce uncertainty to the LCA-results as described in section 3.3.1.2. Therefore, a novel analytic approach to analyze those uncertainties due to multi-functionality problems is derived in the following chapter.

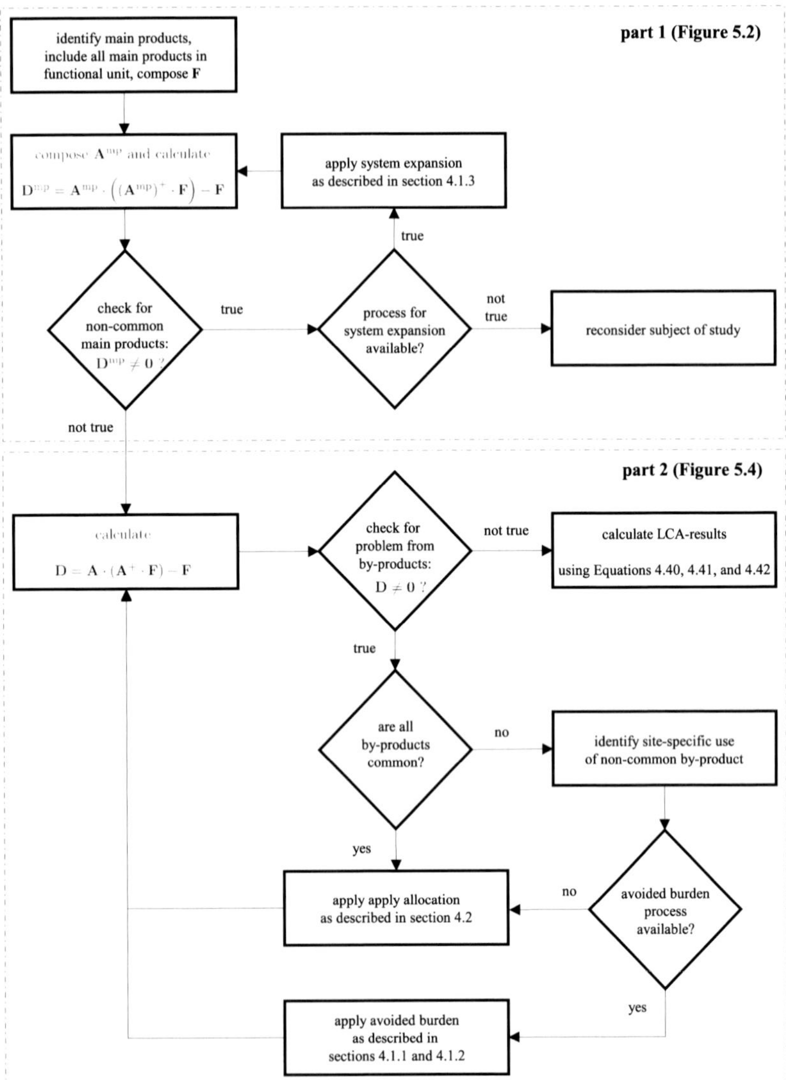

Figure 5.5: Procedure for fixing multi-functionality problems in comparative LCA: two-part workflow for identifying and fixing multi-functionality problems due to non-common main products (part 1) and non-common by-products (part 2).

Chapter 6

Analysis of uncertainties due to fixing multi-functionality problems

Fixing multi-functionality problems is subject to uncertainties (section 3.3.1.2). These uncertainties are introduced by applying system expansion, avoided burden, or allocation (Table 3.2).

The uncertainties due to fixing multi-functionality problems can be related to parameter uncertainty: uncertainties due to system expansion and avoided burden can be expressed by the uncertainty in aggregation factors; uncertainties due to allocation can be expressed by the uncertainty in allocation factors (cf. chapter 4).

In this chapter, a method is presented for analytical analysis of uncertainties due to fixing multi-functionality problems. The method is based on the formulas derived in chapter 4, i.e., Equations 4.40 to 4.42 on p. 61. The method applies error propagation to uncertainties in aggregation and allocation factors. Earlier versions of this approach were presented (Jung et al., 2012) and submitted for publication (Jung et al., 2013b).

The chapter is organized in three parts. In section 6.1.4, an analytical approach for sensitivity analysis with respect to fixing multi-functionality problems is proposed. For this purpose, sensitivity coefficients of LCA-results are derived with respect to aggregation factors (section 6.1.1), allocation factors (section 6.1.2), and process data (section 6.1.3).

In the second part of this chapter, analytical error propagation is applied for analysis of uncertainties due to fixing multi-functionality. The general approach is derived in section 6.2.1; it is based on the sensitivity coefficients presented in section 6.1.4 and Equations 4.40 to 4.42. Approaches to model uncertainties in aggregation factors used for system expansion or avoided burden are presented in section 6.2.2. Subsequently, modeling uncertainties in allocation factors is discussed in section 6.2.3.

The chapter concludes with a discussion of the proposed methods in section 6.3. All methods presented in this chapter are applied to an example in chapter 7. Earlier versions of the methods are also discussed in Jung et al. (2012, 2013b).

6.1 Sensitivity of LCA-results with respect to fixing multi-functionality problems

Sensitivity is "the rate of change of an output with respect to variation in an input" for a given model (Morgan and Henrion, 1992). For models described by a mathematical function, an output's sensitivity is a partial derivative of the output variable with respect to an input variable. The matrix formulations presented in section 3.1.3.1 allows deriving sensitivity coefficients of LCA-results: Heijungs (2010) derived sensitivity coefficients for scaling vectors $\mathbf{s}$, total intervention vectors $\mathbf{g}$, and total impact vectors $\mathbf{h}$ with respect to changes in the input matrices $\mathbf{A}$, $\mathbf{B}$, and $\mathbf{Q}$. These coefficients model sensitivities of LCA-results with respect to process data in the matrices $\mathbf{A}$ and $\mathbf{B}$ as well as characterization factors in matrix $\mathbf{Q}$. However, the influence of fixing multi-functionality problems was excluded from the analysis by assuming a square and invertible technology matrix $\mathbf{A}$.

In contrast, this work uses the expanded matrix formulation given in section 4.3 to derive sensitivity coefficients with respect to the parameters for fixing multi-functionality problems. Sensitivity coefficients are derived with respect to aggregation factors in the following section 6.1.1. Coefficients with respect to allocation factors are presented in section 6.1.2. In section 6.1.3, the existing sensitivity coefficient formulas from Heijungs (2010) are adapted to the expanded matrix formulation used in this work. Potential applications of the new coefficients for sensitivity analysis are discussed in section 6.1.4.

The mathematical derivations are based on fundamental matrix algebra (Magnus and Neudecker, 2007; Petersen and Pedersen, 2012). The most important relationship used in this work is the derivative of a $(\Gamma \times \Delta)$-matrix $\mathbf{Z}$ with respect to a matrix element z_{xy}, i.e.,

$$\frac{\partial \mathbf{Z}^{\Gamma \times \Delta}}{\partial z_{xy}} = \mathbf{J}_{xy}^{\Gamma \times \Delta}. \tag{6.1}$$

In Equation 6.1, the $(\Gamma \times \Delta)$-matrix $\mathbf{J}_{xy}$ is a *single-entry-matrix* for which

$$j_{pq} = \begin{cases} 1 \text{ for } p = x \text{ and } q = y \\ 0 \text{ else.} \end{cases} \tag{6.2}$$

The derivatives of a matrix product $\mathbf{RZ}$ and an element-wise matrix product $\mathbf{W} \circ \mathbf{Z}$ are

$$\frac{\partial (\mathbf{R} \cdot \mathbf{Z})}{\partial z_{xy}} = \mathbf{R} \cdot \frac{\partial \mathbf{Z}}{\partial z_{xy}} + \frac{\partial \mathbf{R}}{\partial z_{xy}} \cdot \mathbf{Z}, \tag{6.3}$$

$$\frac{\partial (\mathbf{W} \circ \mathbf{Z})}{\partial z_{xy}} = \mathbf{W} \circ \frac{\partial \mathbf{Z}}{\partial z_{xy}} + \frac{\partial \mathbf{W}}{\partial z_{xy}} \circ \mathbf{Z}. \tag{6.4}$$

In Equations 6.3 and 6.4, it is assumed that a $(\Delta \times \Gamma)$-matrix $\mathbf{R}$ and a $(\Gamma \times \Delta)$-matrix $\mathbf{W}$ are are also functions of the matrix elements z_{xy}. If the matrices $\mathbf{R}$ and $\mathbf{W}$ are independent from the matrix elements z_{xy}, the second term in both Equations 6.3 and 6.4 becomes zero because the partial derivatives are $\frac{\partial \mathbf{R}}{\partial z_{xy}} = \mathbf{0}$ and $\frac{\partial \mathbf{W}}{\partial z_{xy}} = \mathbf{0}$.

Finally, the derivative of an inverted $(\gamma \times \gamma)$-matrix $\mathbf{Z}$ with respect to a matrix element z_{xy} is

$$\frac{\partial \mathbf{Z}^{-1}}{\partial z_{xy}} = -\mathbf{Z}^{-1} \cdot \frac{\partial \mathbf{Z}}{\partial z_{xy}} \cdot \mathbf{Z}^{-1}. \tag{6.5}$$

In the following sections, these rules for matrix derivatives are applied to Equations 4.40 to 4.42 given on p. 61.

Herein, the matrix $\mathbf{A}_{\text{fixed}}$ describes a technology matrix where all multi-functionality problems of an original technology matrix $\mathbf{A}$ are fixed, i.e.,

$$\mathbf{A}_{\text{fixed}} = (\mathbf{U} \circ (\mathbf{EAV} \cdot \mathbf{T}_1)) \cdot \mathbf{C}. \tag{6.6}$$

Aggregation factors used during system expansion or avoided burden and allocation factors are comprised in the matrices $\mathbf{V}$ and $\mathbf{C}$, respectively. The matrix $\mathbf{B}_{\text{fixed}}$ represents accordingly an intervention matrix where all multi-functionality problems of an original intervention matrix $\mathbf{B}$ are fixed, i.e.,

$$\mathbf{B}_{\text{fixed}} = \mathbf{B} \cdot \mathbf{V} \cdot \mathbf{T}_0 \cdot \mathbf{C}. \tag{6.7}$$

6.1.1 Sensitivity coefficients for aggregation factors

In section 4.1.2, the discrete and potentially ambiguous choice between multiple additional or avoided processes is modeled by aggregating multiple processes to a single process. This aggregation is described by using aggregation factors v_{jl} collected in the aggregation matrix $\mathbf{V}$, cf. Equation 4.13 on p. 51. Aggregation factors model the discrete choice between multiple processes for system expansion or avoided burden as continuous parameter. Continuous parameters allow analytical sensitivity analysis based on first-order TAYLOR-approximations, cf. section 3.3.2.3.

Following Equation 6.1, the partial derivative of the $(n \times w)$-aggregation matrix $\mathbf{V}$ with respect to an aggregation factor v_{jl} is

$$\frac{\partial \mathbf{V}}{\partial v_{jl}} = \mathbf{J}_{jl}^{n \times w}. \tag{6.8}$$

Sensitivity coefficients for LCA-results can be obtained by applying the rule for derivatives of matrix products (cf. Equation 6.3) to Equations 4.40 to 4.42. A vector of sensitivity coefficients for a scaling vector $\mathbf{s}$ with respect to an aggregation factor v_{jl} is

$$\begin{aligned} \frac{\partial \mathbf{s}}{\partial v_{jl}} &= -\mathbf{A}_{\text{fixed}}^{-1} \cdot \overbrace{\left[\mathbf{U} \circ \left[\left(\mathbf{EAJ}_{jl}^{n \times w}\right)\mathbf{T}_1\right]\right]\mathbf{C}}^{\frac{\partial \mathbf{A}_{\text{fixed}}}{\partial v_{jl}}} \cdot \overbrace{\mathbf{A}_{\text{fixed}}^{-1} \cdot \mathbf{Ef}}^{\mathbf{s}} \\ &= -\mathbf{A}_{\text{fixed}}^{-1} \cdot \left[\mathbf{U} \circ \left[\left(\mathbf{EAJ}_{jl}^{n \times w}\right)\mathbf{T}_1\right]\right]\mathbf{C} \cdot \mathbf{s}. \end{aligned} \tag{6.9}$$

Calculation the vector of sensitivity coefficient $\frac{\partial \mathbf{s}}{\partial v_{jl}}$ from Equation 6.9 requires inversion of the fixed technology matrix $\mathbf{A}_{\text{fixed}}$ (Equation 6.6). However, this inversion is already necessary for calculation of LCA-results from Equations 4.40 to 4.42. Calculating a vector of sensitivity coefficients does thus not considerably increase the computational effort of calculating only LCA-results.

The corresponding vectors of sensitivity coefficients for the total intervention vector $\mathbf{g}$ and the total impact vector $\mathbf{h}$ with respect to an aggregation factor v_{jl} are

$$\begin{aligned} \frac{\partial \mathbf{g}}{\partial v_{jl}} &= \mathbf{BJ}_{jl}^{n \times w}\mathbf{T}_0\mathbf{Cs} + \mathbf{BVT}_0\mathbf{C} \cdot \frac{\partial \mathbf{s}}{\partial v_{jl}} \\ &= \mathbf{B}\left[\mathbf{J}_{jl}^{n \times w}\mathbf{T}_0 - \mathbf{VT}_0\mathbf{CA}_{\text{fixed}}^{-1}\left[\mathbf{U} \circ \left[\left(\mathbf{EAJ}_{jl}^{n \times w}\right)\mathbf{T}_1\right]\right]\right]\mathbf{Cs}, \end{aligned} \tag{6.10}$$

$$\begin{aligned} \frac{\partial \mathbf{h}}{\partial v_{jl}} &= \mathbf{Q} \cdot \frac{\partial \mathbf{g}}{\partial v_{jl}} \\ &= \mathbf{QB}\left[\mathbf{J}_{jl}^{n\times w}\mathbf{T}_0 - \mathbf{VT}_0\mathbf{CA}_{\text{fixed}}^{-1}\left[\mathbf{U} \circ \left[\left(\mathbf{EAJ}_{jl}^{n\times w}\right)\mathbf{T}_1\right]\right]\right]\mathbf{Cs}. \end{aligned} \tag{6.11}$$

Here, the only difference between Equation 6.10 and 6.11 is the multiplication of the characterization matrix $\mathbf{Q}$ in the latter equation. The formulas in both Equations 6.10 and 6.11 consist of two terms: The two terms origin from applying the product rule for matrix derivatives (Equation 6.3) to Equations 4.41 and 4.42. The two terms stem from aggregation of unit processes in the intervention matrix $\mathbf{B}$ (1st term) and in the technology matrix $\mathbf{A}$ (2nd term). Still, calculating the vectors of sensitivity coefficients from Equations 6.10 and 6.11 requires only one matrix inversion, i.e., inversion of the fixed technology matrix $\mathbf{A}_{\text{fixed}}$.

6.1.2 Sensitivity coefficients for allocation factors

Allocation factors c_{pq} model how environmental impacts of multi-functional processes are allocated to mono-functional processes. In the matrix formulation derived in chapter 4, these allocation factors are collected in an allocation matrix $\mathbf{C}$, cf. Equation 4.31 on page 58. Sensitivity coefficients for LCA-results with respect to allocation factors are obtained from Equations 4.40 to 4.42.

The partial derivative of an $((w+\beta)\times r)$-allocation matrix $\mathbf{C}$ with respect to an allocation factor c_{pq} is

$$\frac{\partial \mathbf{C}}{\partial c_{pq}} = \mathbf{J}_{pq}^{(w+\beta)\times r}. \tag{6.12}$$

Applying the rules for derivatives of matrix products (cf. Equations 6.3 and 6.4) yields a vector of sensitivity coefficients for a scaling vector $\mathbf{s}$ with respect to an allocation factor c_{pq}, i.e.,

$$\begin{aligned} \frac{\partial \mathbf{s}}{\partial c_{pq}} &= -\mathbf{A}_{\text{fixed}}^{-1} \cdot \overbrace{\left[\mathbf{U} \circ \left[\left(\mathbf{EAV}\right)\mathbf{T}_1\right]\right]\mathbf{J}_{pq}^{(w+\beta)\times r}}^{\frac{\partial \mathbf{A}_{\text{fixed}}}{\partial c_{pq}}} \cdot \overbrace{\mathbf{A}_{\text{fixed}}^{-1} \cdot \mathbf{Ef}}^{\mathbf{s}} \\ &= -\mathbf{A}_{\text{fixed}}^{-1} \cdot \left[\mathbf{U} \circ \left[\left(\mathbf{EAV}\right)\mathbf{T}_1\right]\right]\mathbf{J}_{pq}^{(w+\beta)\times r} \cdot \mathbf{s} \end{aligned} \tag{6.13}$$

As in Equation 6.9, the vector of sensitivity coefficients $\frac{\partial \mathbf{s}}{\partial c_{pq}}$ is calculated from Equation 6.13 using only one matrix inversion: The inversion of the fixed technology matrix

A is already required for calculating LCA-results (Equations 4.40 to 4.42). Again, the computation effort for calculating the vector of sensitivity coefficients should remain in the order of calculating LCA-results.

The vectors of sensitivity coefficients for the total intervention vector **g** and the total impact vector **h** with respect to an allocation factor c_{pq} are

$$\begin{aligned}
\frac{\partial \mathbf{g}}{\partial c_{pq}} &= \mathbf{BVT}_0\mathbf{J}_{pq}^{(w+\beta)\times r}\mathbf{s} + \mathbf{BVT}_0\mathbf{C}\cdot\frac{\partial \mathbf{s}}{\partial c_{pq}} \\
&= \mathbf{BVT}_0\left[\mathbf{I}^{(w+\beta)\times(w+\beta)} - \mathbf{CA}_{\text{fixed}}^{-1}\left[\mathbf{U}\circ\left[(\mathbf{EAV})\,\mathbf{T}_1\right]\right]\right]\cdot\mathbf{J}_{pq}^{(w+\beta)\times r}\cdot\mathbf{s}, \quad (6.14)\\
\frac{\partial \mathbf{h}}{\partial c_{pq}} &= \mathbf{Q}\cdot\frac{\partial \mathbf{g}}{\partial c_{pq}} \\
&= \mathbf{QBVT}_0\left[\mathbf{I}^{(w+\beta)\times(w+\beta)} - \mathbf{CA}_{\text{fixed}}^{-1}\left[\mathbf{U}\circ\left[(\mathbf{EAV})\,\mathbf{T}_1\right]\right]\right]\cdot\mathbf{J}_{pq}^{(w+\beta)\times r}\cdot\mathbf{s} \quad (6.15)
\end{aligned}$$

Here, the matrix **I** is a $((w+\beta)\times(w+\beta))$-identity matrix in Equations 6.14 and 6.15.

The structure of the formulas in Equations 6.14 and 6.15 is similar to that of Equations 6.10 and 6.11: The two terms result from the product rule for matrix derivatives (Equation 6.3).

6.1.3 Sensitivity coefficients for process data and characterization factors

Equations 4.40 to 4.42 also allow deriving sensitivity coefficients for LCA-results with respect to process data and characterization factors. In matrix-based LCA, process data is collected economic flows in the technology matrix **A** and as elementary flows the intervention matrix **B**. Characterization factors are comprised in the characterization matrix **Q**.

Vectors of sensitivity coefficients for a scaling vector **s**, a total intervention vector **g**, and a total impact vector **h** with respect to a technology matrix element a_{ij} are

$$\frac{\partial \mathbf{s}}{\partial a_{ij}} = -\mathbf{A}_{\text{fixed}}^{-1}\left[\mathbf{U}\circ\left[\left(\mathbf{EJ}_{ij}^{m\times n}\mathbf{V}\right)\mathbf{T}_1\right]\mathbf{C}\right]\cdot\mathbf{s}, \tag{6.16}$$

$$\frac{\partial \mathbf{g}}{\partial a_{ij}} = -\mathbf{B}_{\text{fixed}}\cdot\mathbf{A}_{\text{fixed}}^{-1}\left[\mathbf{U}\circ\left[\left(\mathbf{EJ}_{ij}^{m\times n}\mathbf{V}\right)\mathbf{T}_1\right]\mathbf{C}\right]\cdot\mathbf{s}, \tag{6.17}$$

$$\frac{\partial \mathbf{h}}{\partial a_{ij}} = -\mathbf{Q}\cdot\mathbf{B}_{\text{fixed}}\cdot\mathbf{A}_{\text{fixed}}^{-1}\left[\mathbf{U}\circ\left[\left(\mathbf{EJ}_{ij}^{m\times n}\mathbf{V}\right)\mathbf{T}_1\right]\mathbf{C}\right]\cdot\mathbf{s}. \tag{6.18}$$

Intervention matrix elements b_{ij}, i.e. the elementary flows, influence the total intervention vector $\mathbf{g}$ and the total impact vector $\mathbf{h}$, cf. Equations 4.40 to 4.42. The vectors of the corresponding sensitivity coefficients with respect to an elementary flow b_{ij} are

$$\frac{\partial \mathbf{g}}{\partial b_{ij}} = \mathbf{J}_{ij}^{v \times n} \cdot \mathbf{V}\mathbf{T}_0\mathbf{C}\mathbf{s}, \tag{6.19}$$

$$\frac{\partial \mathbf{h}}{\partial b_{ij}} = \mathbf{Q} \cdot \mathbf{J}_{ij}^{v \times n} \cdot \mathbf{V}\mathbf{T}_0\mathbf{C}\mathbf{s}. \tag{6.20}$$

Characterization matrix elements q_{ki}, i.e. characterization factors, only linked to LCIA-results $\mathbf{h}$. The vector of sensitivity coefficients for a total impact vector $\mathbf{h}$ with respect to a characterization factor q_{ki} is

$$\frac{\partial \mathbf{h}}{\partial q_{ki}} = \mathbf{J}_{ki}^{u \times v} \cdot \mathbf{g}. \tag{6.21}$$

Equations 6.16 to 6.21 are identical with those of Heijungs (2010) if the matrices **E**, **V**, **T**, **U**, and **C** are replaced by identity matrices (sections 4.1 and 4.2). All sensitivity coefficients derived in this work are summarized in Table 6.1.

6.1.4 Sensitivity analysis for fixing multi-functionality problems

The formulas derived in the previous sections calculate absolute sensitivity coefficients. These absolute sensitivity coefficients can be unsuitable for sensitivity analysis because they depend on the absolute values of the inputs. In LCA, process data and characterization factors can be given in varying units and different scales. But sensitivity analysis should be "unaffected by the units of measurement" (Morgan and Henrion, 1992).

This drawback is overcome by normalizing the changes in input and output data as a fraction of their nominal value. A normalized sensitivity coefficient is defined as the ratio of relative change in an output to a relative change in an input: For a given function $f(x)$ and an input data x_0, the normalized sensitivity coefficient[1] is

$$\frac{\partial f(x)/f(x_0)}{\partial x/x_0}. \tag{6.22}$$

[1] In Heijungs (2010), normalized sensitivity coefficients are called multipliers.

Table 6.1: Sensitivity coefficients for scaling vector **s**, total intervention vector **g**, and total impact vector **h** with respect to process data a_{ij} and b_{ij}, characterization factors q_{ki}, aggregation factors v_{jl}, and allocation factors c_{pq} for matrix-based LCA

	$\frac{\partial \mathbf{s}}{\cdots}$	$\frac{\partial \mathbf{g}}{\cdots}$	$\frac{\partial \mathbf{h}}{\cdots}$
$\frac{\cdots}{\partial a_{ij}}$	$-\mathbf{A}_{\text{fixed}}^{-1}\left[\mathbf{U}\circ\left[\left(\mathbf{E}\mathbf{J}_{ij}^{m\times n}\mathbf{V}\right)\mathbf{T}_1\right]\mathbf{C}\right]\cdot\mathbf{s}$	$\mathbf{B}_{\text{fixed}}\cdot\frac{\partial \mathbf{s}}{\partial a_{ij}}$	$\mathbf{Q}\cdot\frac{\partial \mathbf{g}}{\partial a_{ij}}$
$\frac{\cdots}{\partial b_{ij}}$	0	$\mathbf{J}_{ij}^{v\times n}\cdot\mathbf{V}\mathbf{T}_0\mathbf{C}\mathbf{s}$	$\mathbf{Q}\cdot\frac{\partial \mathbf{g}}{\partial b_{ij}}$
$\frac{\cdots}{\partial q_{ki}}$	0	0	$\mathbf{J}_{ki}^{u\times v}\cdot\mathbf{g}$
$\frac{\cdots}{\partial v_{jl}}$	$-\mathbf{A}_{\text{fixed}}^{-1}\cdot\left[\mathbf{U}\circ\left[\left(\mathbf{E}\mathbf{A}\mathbf{J}_{jl}^{n\times w}\right)\mathbf{T}_1\right]\right]\cdot\mathbf{C}\cdot\mathbf{s}$	$\mathbf{B}\left[\mathbf{J}_{jl}^{n\times w}\mathbf{T}_0-\mathbf{V}\mathbf{T}_0\mathbf{C}\mathbf{A}_{\text{fixed}}^{-1}\left[\mathbf{U}\circ\left[\left(\mathbf{E}\mathbf{A}\mathbf{J}_{jl}^{n\times w}\right)\mathbf{T}_1\right]\right]\right]\mathbf{C}\mathbf{s}$	$\mathbf{Q}\cdot\frac{\partial \mathbf{g}}{\partial v_{jl}}$
$\frac{\cdots}{\partial c_{pq}}$	$-\mathbf{A}_{\text{fixed}}^{-1}\cdot\left[\mathbf{U}\circ\left[\left(\mathbf{E}\mathbf{A}\mathbf{V}\right)\mathbf{T}_1\right]\right]\cdot\mathbf{J}_{pq}^{(w+\beta)\times r}\cdot\mathbf{s}$	$\mathbf{B}\mathbf{V}\mathbf{T}_0\left[\mathbf{I}^{(w+\beta)\times(w+\beta)}-\mathbf{C}\mathbf{A}_{\text{fixed}}^{-1}\left[\mathbf{U}\circ\left[\left(\mathbf{E}\mathbf{A}\mathbf{V}\right)\mathbf{T}_1\right]\right]\right]\mathbf{J}_{pq}^{(w+\beta)\times r}\mathbf{s}$	$\mathbf{Q}\cdot\frac{\partial \mathbf{g}}{\partial c_{pq}}$

The numerical value of a normalized sensitivity coefficient gives the percent change in $f(x_0)$ caused by an 1 % change of x_0.

Sakai and Yokoyama (2002) and Heijungs (2010) provided explicit formulas for normalized sensitivity coefficients in matrix-based LCA. These authors assumed multi-functionality problems to be fixed by using a square technology matrix $\mathbf{A}$. In this work, additional normalized sensitivity coefficients for aggregation and allocation factors allow sensitivity analysis of aggregation and allocation factors. Following the notation of Heijungs (2010), the normalized sensitivity coefficient of the qth scaling factor s_q is written σ_q. The normalized sensitivity coefficients γ_i and η_k refer accordingly to sensitivity coefficients of the ith total intervention vector element g_i and the kth total impact vector element h_k.

Table 6.2 summarizes the normalized sensitivity coefficients with respect to process data a_{ij} and b_{ij}, characterization factors q_{ki}, as well as aggregation factors v_{jl} and allocation factors c_{pq}.

Table 6.2: Normalized sensitivity coefficients $\sigma_i(\cdots)$, $\gamma_i(\cdots)$, and $\eta_k(\cdots)$ for the scaling vector element s_q, the total intervention vector element g_i, and the total impact vector element h_k with respect to process data a_{ij} and b_{ij}, characterization factors q_{ki}, aggregation factors v_{jl}, and allocation factors c_{pq}

coefficient	a_{ij}	b_{ij}	q_{ki}	v_{jl}	c_{pq}
$\sigma_q(\cdots)$	$\frac{\partial s_q}{\partial a_{ij}} \cdot \frac{a_{ij}}{s_q}$	$\frac{\partial s_q}{\partial b_{ij}} \cdot \frac{b_{ij}}{s_q}$	$\frac{\partial s_q}{\partial q_{ki}} \cdot \frac{q_{ki}}{s_q}$	$\frac{\partial s_q}{\partial v_{jl}} \cdot \frac{v_{jl}}{s_q}$	$\frac{\partial s_q}{\partial c_{pq}} \cdot \frac{c_{pq}}{s_q}$
$\gamma_i(\cdots)$	$\frac{\partial g_i}{\partial a_{ij}} \cdot \frac{a_{ij}}{g_i}$	$\frac{\partial g_i}{\partial b_{ij}} \cdot \frac{b_{ij}}{g_i}$	$\frac{\partial g_i}{\partial q_{ki}} \cdot \frac{q_{ki}}{g_i}$	$\frac{\partial g_i}{\partial v_{jl}} \cdot \frac{v_{jl}}{g_i}$	$\frac{\partial g_i}{\partial c_{pq}} \cdot \frac{c_{pq}}{g_i}$
$\eta_k(\cdots)$	$\frac{\partial h_k}{\partial a_{ij}} \cdot \frac{a_{ij}}{h_k}$	$\frac{\partial h_k}{\partial b_{ij}} \cdot \frac{b_{ij}}{h_k}$	$\frac{\partial h_k}{\partial q_{ki}} \cdot \frac{q_{ki}}{h_k}$	$\frac{\partial h_k}{\partial v_{jl}} \cdot \frac{v_{jl}}{h_k}$	$\frac{\partial h_k}{\partial c_{pq}} \cdot \frac{c_{pq}}{h_k}$

Normalized sensitivity coefficients are systematic measures for identifying influential parameters within a process system. Heijungs (2010) applied normalized sensitivity coefficients to process data a_{ij} and b_{ij}, and characterization factors q_{ki}. The new coefficients derived in this work additionally integrate aggregation factors v_{jl} and allocation factors c_{pq}. The equations summarized in Table 6.2 allow comparing the influence of aggregation and allocation factors with that of process data and characterization factors. Still, uncertainties in all those parameters are not yet considered. Parameters with low sensitivity coefficients can still introduce large uncertainties if the parameter is uncertain itself. In the following section 6.2, a method for uncertainty analysis is derived which considers all those uncertainties.

6.2 Uncertainty of LCA-results due to fixing multi-functionality problems

6.2.1 Propagation of uncertainties due to fixing multi-functionality problems

An analytical approach for propagating uncertainties in process data and characterization factors is described in section 3.3.2.3. In this section, that analytical approach is further developed to propagating uncertainties due to fixing multi-functionality problems (Table 3.2). For this purpose, analytical error propagation is applied to Equations 4.42 derived in section 4.3. Thereby, uncertainty in process data, characterization factors, aggregation factors, and allocation factors is propagated to uncertainty in the total impact vector **h**.

Following Equation 4.42, the total impact vector **h** of a multi-functional process system is calculated from

$$\begin{aligned} \mathbf{h} &= \mathbf{h}(\mathbf{Q}, \mathbf{B}, \mathbf{V}, \mathbf{T}_0, \mathbf{T}_1, \mathbf{U}, \mathbf{E}, \mathbf{A}, \mathbf{C}, \mathbf{f}) \\ &= \mathbf{Q} \cdot \mathbf{B}\mathbf{V}\mathbf{T}_0\mathbf{C} \cdot \left(\left[\mathbf{U} \circ \left((\mathbf{E}\mathbf{A}\mathbf{V}) \, \mathbf{T}_1 \right) \right] \mathbf{C} \right)^{-1} \cdot \mathbf{E}\mathbf{f}. \end{aligned} \tag{6.23}$$

For a given function $f(x_i) = f(x_1, \ldots, x_n)$, the total variance of the function output $\mathrm{var} f(x_i)$ is approximated from n input variances $\mathrm{var}(x_i)$, i.e.,

$$\mathrm{var}(f) \approx \sum_{i=1}^{n} \left(\frac{\partial f}{\partial x_i} \right)^2 \mathrm{var}(x_i) + 2 \sum_{i=1}^{n} \sum_{j=i+1}^{n} \frac{\partial f}{\partial x_i} \frac{\partial f}{\partial x_j} \mathrm{cov}(x_i, x_j). \tag{6.24}$$

In Equation 6.23, the transformation matrices $\mathbf{T}_0$ and $\mathbf{T}_1$, the function matrix **U**, the equivalence matrix **E**, and the final demand vector **f** are specified by the LCA-practitioner. Herein, the zero-one pattern of the matrices $\mathbf{T}_0$, $\mathbf{T}_1$, **U**, and **E** is predetermined by the multi-functional unit processes. The final demand vector **f** is chosen to represent the goal of a LCA study. Therefore, the matrices $\mathbf{T}_0$, $\mathbf{T}_1$, **U**, **E**, and the final demand vector **f** are not subject to uncertainties.

In contrast, process data a_{ij} and b_{ij}, characterization factors q_{ki}, aggregation factors v_{jl}, and allocation factors c_{pq} are subject to uncertainties which propagate to the total impact vector $\mathbf{h}$. The propagation of these uncertainties can be modeled by applying Equation 3.31 to Equation 6.23: the variance of the kth total impact vector element h_k is thus approximated from the variances of economic flows a_{ij}, elementary flows b_{ij}, characterization factors q_{ki}, aggregation factors v_{jl}, and allocation factors c_{pq}, i.e.,

$$\begin{aligned}
\mathrm{var}(h_k) \approx\; & \overbrace{\sum_{i,j}\left(\frac{\partial h_k}{\partial a_{ij}}\right)^2 \mathrm{var}(a_{ij}) + 2\sum_{i,j}\sum_{\substack{i'=i+1\\ j'=j+1}} \frac{\partial h_k}{\partial a_{ij}}\frac{\partial h_k}{\partial a_{i'j'}}\mathrm{cov}(a_{ij}, a_{i'j'})^{\to 0}}^{\text{process data: economic flows}} + \\
& \overbrace{\sum_{i,j}\left(\frac{\partial h_k}{\partial b_{ij}}\right)^2 \mathrm{var}(b_{ij}) + 2\sum_{i,j}\sum_{\substack{i'=i+1\\ j'=j+1}} \frac{\partial h_k}{\partial b_{ij}}\frac{\partial h_k}{\partial b_{i'j'}}\mathrm{cov}(b_{ij}, b_{i'j'})^{\to 0}}^{\text{process data: elementary flows}} + \\
& \overbrace{\sum_{i}\left(\frac{\partial h_k}{\partial q_{ki}}\right)^2 \mathrm{var}(q_{ki}) + 2\sum_{i}\sum_{i'=i+1} \frac{\partial h_k}{\partial q_{ki}}\frac{\partial h_k}{\partial q_{ki'}}\mathrm{cov}(q_{ki}, q_{ki'})^{\to 0}}^{\text{impact assessment: characterization factors}} + \\
& \overbrace{\sum_{j,l}\left(\frac{\partial h_k}{\partial v_{jl}}\right)^2 \mathrm{var}(v_{jl}) + 2\sum_{j,l}\sum_{j'=j+1} \frac{\partial h_k}{\partial v_{jl}}\frac{\partial h_k}{\partial v_{j'l}}\underbrace{\mathrm{cov}(v_{jl}, v_{j'l})}_{\substack{\text{for all } l:\\ \sum_{j=1}^{n} v_{jl}=1}}}^{\text{system expansion/avoided burden: aggregation factors}} + \\
& \overbrace{\sum_{p,q}\left(\frac{\partial h_k}{\partial c_{pq}}\right)^2 \mathrm{var}(c_{pq}) + 2\sum_{p,q}\sum_{q'=q+1} \frac{\partial h_k}{\partial c_{pq}}\frac{\partial h_k}{\partial c_{pq'}}\underbrace{\mathrm{cov}(c_{pq}, c_{pq'})}_{\substack{\text{for all } p:\\ \sum_{q=1}^{r} c_{pq}=1}}}^{\text{allocation: alloctation factors}} . \qquad (6.25)
\end{aligned}$$

Here, correlations between the matrix elements a_{ij}, b_{ij}, q_{ki}, and the matrix elements v_{jl} and c_{pq} are also neglected and not shown in Equation 6.25.

In Equation 6.25, it is assumed that all economic flows a_{ij}, all elementary flows b_{ij}, and all characterization factors q_{ki} are uncorrelated. The corresponding covariances $\mathrm{cov}(a_{ij}, a_{i'j'})$, $\mathrm{cov}(b_{ij}, b_{i'j'})$, and $\mathrm{cov}(q_{ki}, q_{ki'})$ are thus canceled to zero. In LCA, unit processes are usually modeled independently. Therefore, this assumption is accurate for the ith economic or elementary flow from different unit processes j and $j+1$, e.g., for $\mathrm{cov}(a_{ij}, a_{i(j+1)})$ or $\mathrm{cov}(b_{ij}, b_{i(j+1)})$.

For some ith and $(i+1)$ economic or elementary flows of the jth unit process, the assumption is inaccurate, because these process data likely correlate, e.g., CO_2-emissions from the electricity generation process correlate with the demand for coal.

These correlations as well as correlations of characterization factors q_{ki} are mostly unavailable for unit processes in LCA (Heijungs, 2010). Therefore, this work follows Heijungs (2010) in neglecting covariances of economic flows a_{ij}, elementary flows b_{ij}, and characterization factors q_{ki}.

In contrast, the correlation of aggregation factors and the correlation the of allocation factors is included in the presented approach. Aggregation factors correlate, because the sum of aggregation factors is constant, cf. Equation 4.12. In a similar way, allocation factors are correlated as well, because their sum is also constant, cf. Equation 4.33.

With these assumptions, Equation 6.25 is simplified to

$$\begin{aligned}\mathrm{var}(h_k) \approx\ & \overbrace{\sum_{i,j}\left(\left(\frac{\partial h_k}{\partial a_{ij}}\right)^2 \mathrm{var}(a_{ij}) + \left(\frac{\partial h_k}{\partial b_{ij}}\right)^2 \mathrm{var}(b_{ij}) + \left(\frac{\partial h_k}{\partial q_{ki}}\right)^2 \mathrm{var}(q_{ki})\right)}^{\text{uncertainty in uncorrelated process data and characterization factors}} \\ & \overbrace{+\sum_{j,l}\left(\frac{\partial h_k}{\partial v_{jl}}\right)^2 \mathrm{var}(v_{jl}) + 2\sum_{j,l}\sum_{j'=j+1}\frac{\partial h_k}{\partial v_{jl}}\frac{\partial h_k}{\partial v_{j'l}}\mathrm{cov}(v_{jl}, v_{j'l})}^{\text{uncertainty correlated aggregation factors}} \\ & \overbrace{+\sum_{p,q}\left(\frac{\partial h_k}{\partial c_{pq}}\right)^2 \mathrm{var}(c_{pq}) + 2\sum_{p,q}\sum_{q'=q+1}\frac{\partial h_k}{\partial c_{pq}}\frac{\partial h_k}{\partial c_{pq'}}\mathrm{cov}(c_{pq}, c_{pq'})}^{\text{uncertainty correlating allocation factors}}. \end{aligned} \tag{6.26}$$

In Equation 6.26, the total variance $\mathrm{var}(h_k)$ includes contributions from uncorrelated process data a_{ij} and b_{ij} and uncorrelated characterization factors q_{ki}. These contributions were also described earlier (cf. Heijungs (2010) in Table 5). Equation 6.26 additionally includes contributions from correlated aggregation factors v_{jl} and correlated allocation factors c_{pq} used for fixing multi-functionally problems in LCA.

Equation 6.26 is a novel approach for holistic uncertainty analysis in LCA. It allows quantifying uncertainty contributions from uncertain process data, uncertain characterization factors, and from uncertainties in fixing multi-functionality problems in a single method. The relative contribution ζ of uncertainty in process data a_{ij} and b_{ij} to the total uncertainty in the LCIA-result h_k is

$$\zeta(h_k, \mathbf{A}, \mathbf{B}) = \frac{\sum_{i,j}\left(\frac{\partial h_k}{\partial a_{ij}}\right)^2 \mathrm{var}(a_{ij}) + \left(\frac{\partial h_k}{\partial b_{ij}}\right)^2 \mathrm{var}(b_{ij})}{\mathrm{var}(h_k)}. \tag{6.27}$$

The relative contributions of the uncertainty in characterization factors q_{ki}, aggregation factors v_{jl}, and allocation factors c_{pq} are,

$$\zeta(h_k, \mathbf{Q}) = \frac{\sum_i \left(\frac{\partial h_k}{\partial q_{ki}}\right)^2 \operatorname{var}(q_{ki})}{\operatorname{var}(h_k)}, \tag{6.28}$$

$$\zeta(h_k, \mathbf{V}) = \frac{\sum_{j,l} \left(\frac{\partial h_k}{\partial v_{jl}}\right)^2 \operatorname{var}(v_{jl}) + 2\sum_{j,l}\sum_{j'=j+1} \frac{\partial h_k}{\partial v_{jl}} \frac{\partial h_k}{\partial v_{j'l}} \operatorname{cov}(v_{jl}, v_{j'l})}{\operatorname{var}(h_k)}, \tag{6.29}$$

and

$$\zeta(h_k, \mathbf{C}) = \frac{\sum_{p,q} \left(\frac{\partial h_k}{\partial c_{pq}}\right)^2 \operatorname{var}(c_{pq}) + 2\sum_{p,q}\sum_{q'=q+1} \frac{\partial h_k}{\partial c_{pq}} \frac{\partial h_k}{\partial c_{pq'}} \operatorname{cov}(c_{pq}, c_{pq'})}{\operatorname{var}(h_k)}. \tag{6.30}$$

Evaluating Equation 6.26 requires partial derivatives of the total impact vector $\mathbf{h}$ with respect to a_{ij}, b_{ij}, q_{ki}, v_{jl}, and c_{pq} as well as variance and covariance data of these parameters. These partial derivatives were derived in section 6.1 and summarized in Table 6.1 on page 90.

Given that sufficient data is available, determining variances of process data a_{ij} and b_{ij} and characterization factors q_{ki} follows statistical methods.[2]

The following sections 6.2.2 and 6.2.3 present methods for determining variance and covariance data for aggregation factors v_{jl} and allocation factors c_{pq}.

6.2.2 Modeling uncertainties due to system expansion and avoided burden

System expansion and avoided burden require added or avoided processes to fix multi-functionality problems in comparative LCA sections 3.2.3.1 and 3.2.3.2). In section 4.1.2, aggregation factors are introduced to model the potentially ambiguous choice between multiple candidates for added or avoided processes. Two approaches for selecting these processes exist for LCA-practitioners, cf. Sonnemann and Vigon (2011) on p. 79 and European Commission (2010b) on p. 87ff: discrete marginal processes or mix (average) processes. Both approaches are subject to uncertainties

[2] An example for available variance data is the ecoinvent database 2.0, cf. Frischknecht et al. (2005).

(Table 3.2). In this work, those uncertainties are modeled using uncertain aggregation factors.

In section 6.2.2.1, a model for uncertainties due to using mix processes with constant mix compositions is suggested. An approach for modeling the uncertainty due to the choice between discrete marginal processes is proposed in section 6.2.2.2. The models for both uncertainties can be simplified for the choice between two discrete marginal processes or a mix process composed of two processes. This simplification for the aggregation of two unit processes is described in section 6.2.2.3. A model for the uncertainties due to additional process data from added and avoided processes is modeled in section 6.2.2.4.

6.2.2.1 Modeling uncertainties due to mix processes with constant mix composition

Mix processes are recommended to be used as added or avoided process (Sonnemann and Vigon (2011) on p. 79 and European Commission (2010b) on pp. 88-89). Those mix processes are typically provided with a constant mix composition, e.g., the German electricity mix for a given year. Using a mix process with a constant mix composition is a simplification that is subject to uncertainties: The composition of mix process changes over time (e.g., electricity mix in Germany). Those uncertainties can be modeled as errors in aggregation factors that represent a mix composition (cf. section 4.1.1 on pp. 49-50).

The method presented in section 6.2 allows propagating errors in aggregation factors into LCA-results. Thereby, the contribution of uncertainty due to uncertain aggregation factors can be quantified and compared to other sources of uncertainty in LCA-results (Equation 6.26). For evaluation of Equation 6.26, variance and covariance data of aggregation factors v_{jl} have to be estimated. In this section, a method is proposed for modeling variance and covariance data in aggregation factors.

For this purpose, it is suggest to calculate an average aggregation factor $\tilde{v}_{jl}$ from a total number of Ω aggregation factors v_{jl}^{ω}, i.e.,

$$\tilde{v}_{jl} = \frac{1}{\Omega} \sum_{\omega=1}^{\Omega} v_{jl}^{\omega}. \tag{6.31}$$

Each set of aggregation factors v_{jl}^{ω} represents a defined mix composition, e.g., taken from historic mix compositions. For many mix processes, the historic development of the mix compositions are known and published by authorities (e.g., federal statistical offices), trade associations (e.g., associations of energy suppliers such as Bundesverband Energie- und Wasserwirtschaft (2013)), or historic datasets of LCI-databases such as ecoinvent (Frischknecht et al., 2005).

In this work, it is proposed to model the uncertainty due to using a constant mix composition by calculating the unbiased sample variance and covariance for the average aggregation factor $\tilde{v}_{jl}$ (cf. Morgan and Henrion (1992) on p. 80). The unbiased sample variance $\mathrm{var}(\tilde{v}_{jl})$ is calculated from

$$\mathrm{var}(\tilde{v}_{jl}) = \frac{1}{\Omega - 1} \sum_{\omega=1}^{\Omega} \left(v_{jl}^{\omega} - \tilde{v}_{jl}\right)^2. \tag{6.32}$$

The unbiased sample covariance of two average aggregation factors $\tilde{v}_{jl}$ and $\tilde{v}_{j'l}$ becomes

$$\mathrm{cov}(\tilde{v}_{jl}, \tilde{v}_{j'l}) = \frac{1}{\Omega - 1} \sum_{\omega=1}^{\Omega} \left(v_{jl}^{\omega} - \tilde{v}_{jl}\right) \left(v_{j'l}^{\omega} - \tilde{v}_{j'l}\right). \tag{6.33}$$

In Equations 6.32 and 6.33, the unbiased sample variance and covariance is preferred over the biased variance and covariance because the average aggregation factors used for calculation is an estimate itself based on sets of historic mix compositions.

The variance and covariance calculated from Equations 6.32 and 6.33 can be used in Equation 6.26 to analyze the uncertainty due to using a mix process with a constant mix composition. The approach is applied to an example in section 7.3.1.

6.2.2.2 Modeling uncertainties due to selection of marginal process

Using a single marginal processes for system expansion or avoided burden leads to potentially ambiguous choices between multiple candidates for those marginal processes. Such a choice exists if the marginal process is not known and a total number of α processes are considered as candidates for the marginal process. In this work, the uncertainty due to selecting the marginal process is modeled with an uncertain average aggregation factor for aggregating candidate processes for the marginal process.

If process j is used as added or avoided process, the corresponding aggregation factor is $v_{jl} = 1$ and the aggregation factors of other candidate processes are $v_{(j+1)l} = v_{(j+\alpha-1)l} = 0$. Thereby, the condition for the sum of aggregation factors is fulfilled (cf. Equation 4.12). For a total number of α candidate processes, a total number of α sets of aggregation factors exist: In each set, one aggregation factor is one while all others are zero. Consequently, the average aggregation factors $\tilde{v}_{jl}, \ldots \tilde{v}_{(j+\alpha-1)l}$

calculated from those sets depend only on the total number α of candidate processes, respectively, i.e.,

$$\tilde{v}_{jl} = \frac{1}{\alpha}\sum_{\omega=1}^{\alpha} v_{jl}^{\omega} = \frac{1}{\alpha} = \tilde{v}_{(j+1)l} = \ldots = \tilde{v}_{(j+\alpha-1)l}. \tag{6.34}$$

The error of using such an average aggregation factor $\tilde{v}_{jl}$ can be estimated using the biased sample variance, i.e.,

$$\begin{aligned} \mathrm{var}(\tilde{v}_{jl}) &= \frac{1}{\alpha}\sum_{\omega=1}^{\alpha}\left(v_{jl}^{\omega} - \tilde{v}_{jl}\right)^2 \\ &= \frac{1}{\alpha}\left[\left(1-\frac{1}{\alpha}\right)^2 + \sum_{\omega=1}^{\alpha-1}\left(0-\frac{1}{\alpha}\right)^2\right] \\ &= \frac{\alpha-1}{\alpha^2} = \mathrm{var}(\tilde{v}_{(j+1)l}) = \ldots = \mathrm{var}(\tilde{v}_{(j+\alpha-1)l}). \end{aligned} \tag{6.35}$$

The biased sample covariance of any two average aggregation factors $\tilde{v}_{jl}$ and $\tilde{v}_{(j+1)l}$ becomes

$$\begin{aligned} \mathrm{cov}(\tilde{v}_{jl}, \tilde{v}_{(j+1)l}) &= \frac{1}{\alpha}\sum_{\omega=1}^{\alpha}\left(v_{jl}^{\omega} - \tilde{v}_{jl}\right)\left(v_{(j+1)l}^{\omega} - \tilde{v}_{(j+1)l}\right) \\ &= \frac{1}{\alpha}\left[2(1-\frac{1}{\alpha})(0-\frac{1}{\alpha}) + \sum_{\omega=1}^{\alpha-2}(0-\frac{1}{\alpha})^2\right] \\ &= -\frac{1}{\alpha^2}. \end{aligned} \tag{6.36}$$

Here, it is suggested to use the biased variance and covariance because the calculation of the average aggregation factor itself is not an estimate for a given number of candidate processes.

The variance in Equation 6.35 can also be used as a conservative estimate for the uncertainty due to a mix process with a constant mix composition. In this case, the historic mix compositions represent α sets of aggregation factor each with one aggregation factor being one while all others are zero. Therefore, calculating the unbiased sample variance $(\mathrm{var}(\tilde{v}_{jl}))_{\mathrm{mix,max}}$ represents a conservative estimate for the uncertainty due to a mix process with constant mix composition, i.e.,

$$\begin{aligned}(\text{var}(\tilde{v}_{jl}))_{\text{mix,max}} &= \frac{1}{\alpha-1}\sum_{\omega=1}^{\alpha}\left(v_{jl}^{\omega}-\tilde{v}_{jl}\right)^2 \\ &= \frac{1}{\alpha-1}\left[\left(1-\frac{1}{\alpha}\right)^2+\sum_{\omega=1}^{\alpha-1}\left(0-\frac{1}{\alpha}\right)^2\right] \\ &= \frac{1}{\alpha}. \end{aligned} \tag{6.37}$$

Variances and covariances calculated from Equations 6.35 and 6.36 can be used to estimate the uncertainty due to the choice between multiple candidates for a marginal process used during system expansion or avoided burden. Using the variances and covariances in Equation 6.26 allows calculating the contribution of the uncertainty due to selection of marginal processes to the overall uncertainty of LCA-results. The approach is illustrated in section 7.3.2.

6.2.2.3 An uncertainty model for aggregation of two unit processes

LCA-practitioners are frequently confronted the choice between exactly two candidates for the marginal process, e.g., the old or new boiler in the example discussed in section 5.2 (Figure 5.3). Some important mix processes can also be simplified to mixes of two unit processes, e.g., renewable and non-renewable technologies are aggregated to an electricity generation mix.

The calculation of the variance of a total impact vector element h_k (Equation 6.26) can be simplified for the case of aggregating two unit processes. For this purpose, aggregating two unit processes x and y into a single unit process z using two aggregation factors v_{xz} and v_{yz} is considered in the following.

Assume that a set of Ω aggregation factors v_{xz}^{ω} and v_{yz}^{ω} is given, e.g., from a historic mix composition trend. It then follows from Equation 4.12 that each aggregation factor v_{yz}^{ω} and the corresponding average aggregation factor $\tilde{v}_{yz}$ is

$$v_{yz}^{\omega}=1-v_{xz}^{\omega} \quad \text{and} \quad \tilde{v}_{yz}=1-\tilde{v}_{xz}. \tag{6.38}$$

The unbiased sample variance (cf. Equation 6.32) of the aggregation factors v_{xz}^{ω} and v_{yz}^{ω} are

$$\text{var}(\tilde{v}_{xz})=\frac{1}{\Omega-1}\sum_{\omega=1}^{\Omega}\left(v_{xz}^{\omega}-\tilde{v}_{xz}\right)^2 \tag{6.39}$$

and

$$\begin{aligned}
\mathrm{var}(\tilde{v}_{yz}) &= \mathrm{var}(1-\tilde{v}_{xz}) \\
&= \frac{1}{\Omega-1}\sum_{\omega=1}^{\Omega}\left((1-v_{xz}^{\omega})-(1-\tilde{v}_{xz})\right)^2 \\
&= \frac{1}{\Omega-1}\sum_{\omega=1}^{\Omega}\left(\tilde{v}_{xz}-v_{xz}^{\omega}\right)^2 \\
&= \mathrm{var}(\tilde{v}_{xz}). \qquad (6.40)
\end{aligned}$$

The relationships between $\mathrm{var}(\tilde{v}_{yz})$ and $\mathrm{var}(\tilde{v}_{xz})$ is also valid for biased sample variances: The only difference between the biased and unbiased sample variances is the coefficient of the sum in Equation 6.39. The coefficient is $\frac{1}{\Omega}$ for the biased and $\frac{1}{\Omega-1}$ for the unbiased sample variance.

The covariance of these two aggregation factors v_{xz}^{ω} and v_{yz}^{ω} is calculated from Equation 6.33, i.e.,

$$\begin{aligned}
\mathrm{cov}(\tilde{v}_{xz},\tilde{v}_{yz}) &= \frac{1}{\Omega-1}\sum_{\omega=1}^{\Omega}\left(v_{xz}^{\omega}-\tilde{v}_{xz}\right)\left(v_{yz}^{\omega}-\tilde{v}_{yz}\right) \\
&= \frac{1}{\Omega-1}\sum_{\omega=1}^{\Omega}\left(v_{xz}^{\omega}-\tilde{v}_{xz}\right)\left((1-v_{xz}^{\omega})-(1-\tilde{v}_{xz})\right) \\
&= -\frac{1}{\Omega-1}\sum_{\omega=1}^{\Omega}\left(v_{xz}^{\omega}-\tilde{v}_{xz}\right)^2 \\
&= -\mathrm{var}(\tilde{v}_{xz}). \qquad (6.41)
\end{aligned}$$

With Equations 6.39, 6.40, and 6.41, the uncertainty contribution of aggregation factors in Equation 6.26 can be simplified to

$$\begin{aligned}
\zeta(h_k, v_{xz}, v_{yz})\mathrm{var}(h_k) &= \sum_{j,l}\left(\frac{\partial h_k}{\partial v_{jl}}\right)^2 \mathrm{var}(v_{jl}) + 2\sum_{j,l}\sum_{j'=j+1}\frac{\partial h_k}{\partial v_{jl}}\frac{\partial h_k}{\partial v_{j'l'}}\mathrm{cov}(v_{jl}, v_{j'l'}) \\
&= \left(\frac{\partial h_k}{\partial \tilde{v}_{xz}}\right)^2 \mathrm{var}(\tilde{v}_{xz}) + \left(\frac{\partial h_k}{\partial \tilde{v}_{yz}}\right)^2 \mathrm{var}(\tilde{v}_{yz}) \\
&\quad +2\frac{\partial h_k}{\partial \tilde{v}_{xz}}\frac{\partial h_k}{\partial \tilde{v}_{yz}}\mathrm{cov}(\tilde{v}_{xz},\tilde{v}_{yz})
\end{aligned}$$

$$= \left[\left(\frac{\partial h_k}{\partial \tilde{v}_{xz}} \right)^2 + \left(\frac{\partial h_k}{\partial \tilde{v}_{yz}} \right)^2 - 2 \frac{\partial h_k}{\partial \tilde{v}_{xz}} \frac{\partial h_k}{\partial \tilde{v}_{yz}} \right] \mathrm{var}(\tilde{v}_{xz})$$

$$= \left[\left(\frac{\partial h_k}{\partial \tilde{v}_{xz}} \right) - \left(\frac{\partial h_k}{\partial \tilde{v}_{yz}} \right) \right]^2 \mathrm{var}(\tilde{v}_{xz}). \tag{6.42}$$

The variance of an aggregation factor modeling the selection between candidates for the marginal process depends on the number α of candidate processes (cf. Equations 6.35 and 6.36). For the choice between two candidates for the marginal process, $\alpha = 2$ and the biased sample variance $\mathrm{var}(\tilde{v}_{xz})$ becomes

$$\mathrm{var}(\tilde{v}_{jl}) = \frac{\alpha - 1}{\alpha^2} = 0.25. \tag{6.43}$$

Combining Equations 6.42 and 6.43 yields

$$\zeta(h_k, v_{xz}, v_{yz}) \mathrm{var}(h_k) = 0.25 \left[\left(\frac{\partial h_k}{\partial \tilde{v}_{xz}} \right) - \left(\frac{\partial h_k}{\partial \tilde{v}_{yz}} \right) \right]^2. \tag{6.44}$$

Equation 6.44 can be used as an automated estimator for the uncertainty contribution of the choice between two candidates for a marginal process used as added or avoided process. It could be implemented in software tools and calculated whenever LCA-practitioners have the choice between two candidates for a marginal process. An example for the application of Equation 6.44 is given in section 7.3.2.

6.2.2.4 Modeling uncertainties due to additional process data from added or avoided processes

Avoided burden increases the number of unit processes used for calculating LCA-results. In comparative LCA, system expansion also requires data from the added processes. Thus, both avoided burden and system expansion introduce additional process data uncertainty compared to LCA-results calculated without avoided burden or system expansion. Equation 6.26 allows quantifying the contribution of these additional process data uncertainties.

If all added or avoided processes are denoted by j", then the absolute contribution of process data uncertainty (cf. Equation 6.27) can be split into

$$\zeta(h_k, \mathbf{A}, \mathbf{B}) \mathrm{var}(h_k) = \sum_{i,j} \left(\left(\frac{\partial h_k}{\partial a_{ij}} \right)^2 \mathrm{var}(a_{ij}) + \left(\frac{\partial h_k}{\partial b_{ij}} \right)^2 \mathrm{var}(b_{ij}) \right)$$

$$= \overbrace{\sum_{\substack{i,j\\ j\neq j"}} \left(\left(\frac{\partial h_k}{\partial a_{ij}}\right)^2 \mathrm{var}(a_{ij}) + \left(\frac{\partial h_k}{\partial b_{ij}}\right)^2 \mathrm{var}(b_{ij}) \right)}^{\text{uncertainty in regular process data}}$$
$$\underbrace{+\sum_{\substack{i,j"\\ j"\neq j}} \left(\left(\frac{\partial h_k}{\partial a_{ij"}}\right)^2 \mathrm{var}(a_{ij"}) + \left(\frac{\partial h_k}{\partial b_{ij"}}\right)^2 \mathrm{var}(b_{ij"}) \right)}_{\text{uncertainty in added/avoided process data}} \quad (6.45)$$

In Equation 6.45, uncertainty in regular process data refers to unit processes that are not involved in system expansion or avoided burden. Equation 6.45 is applied to an example in section 7.3.2.

6.2.3 Modeling uncertainties due to allocation

Allocation can be subject to two kinds of uncertainty: Uncertainty due to selection of an allocation criterion and due to economic allocation with constant allocation factors (Table 3.2). In section 4.2, allocation factors are included in explicit formulas for calculating LCA-results Equations 4.40 to 4.42). In this work, it is proposed to model both uncertainties due to allocation by uncertain allocation factors.

In section 6.2.3.1, it is described how uncertainties due to selecting an allocation criterion can be modeled with uncertain allocation factors. In section 6.2.3.2, a method for modeling uncertainties due to constant economic allocation factors is presented. The models proposed in sections 6.2.3.1 and 6.2.3.2 can be simplified for allocation with two functional flows, i.e., for two allocation factors. This simplification is derived in section 6.2.3.3.

6.2.3.1 Modeling uncertainties due to selection of allocation criteria

Selecting an allocation criterion is a potentially ambiguous choice for LCA-practitioners. A choice is required if a total number of K suitable allocation criteria exist. In this work, it is proposed to model the uncertainty due to that choice with an uncertain average allocation factor.

For a total number of K allocation criteria each with a corresponding allocation factor c^{κ}_{pq}, an average allocation factor $\tilde{c}_{pq}$ can be calculated from

$$\tilde{c}_{pq} = \frac{1}{K} \sum_{\kappa=1}^{K} c_{pq}^{\kappa}. \tag{6.46}$$

The error of using such an average allocation factor $\tilde{c}_{pq}$ can be estimated using the unbiased sample variance, i.e.,

$$\mathrm{var}(\tilde{c}_{pq}) = \frac{1}{K-1} \sum_{\kappa=1}^{K} \left(c_{pq}^{\kappa} - \tilde{c}_{pq}\right)^2. \tag{6.47}$$

The unbiased sample variance is preferred over the biased variance because the calculation of an average allocation factor $\tilde{c}_{pq}$ is also an estimation: The allocation criteria used to calculate $\tilde{c}_{pq}$ may not include all possible allocation criteria.

The corresponding sample covariance of two average allocation factors $\tilde{c}_{pq}$ and $\tilde{c}_{pq'}$ becomes

$$\mathrm{cov}(\tilde{c}_{pq}, \tilde{c}_{pq'}) = \frac{1}{K-1} \sum_{\kappa=1}^{K} \left(c_{pq}^{\kappa} - \tilde{c}_{pq}\right) \left(c_{pq'}^{\kappa} - \tilde{c}_{pq'}\right). \tag{6.48}$$

In this work, it is suggested to model the uncertainty due to selection between multiple allocation criteria as uncertainty of an average allocation factor. Thereby, the variances and covariances calculated from Equations 6.47 and 6.48 can be used in Equation 6.26 to propagate the uncertainty due to selection between multiple allocation criteria into uncertainty in LCA-results. This approach is applied to an example in section 7.3.3.

6.2.3.2 Modeling uncertainties due to economic allocation with constant allocation factors

In economic allocation, fluctuating market prices cause fluctuating allocation factors. This fluctuation introduces uncertainty if constant allocation factors are used to calculate LCA-results. The contribution of this uncertainty can be calculated from Equation 6.26. In this work, this uncertainty is also modeled with an uncertain average allocation factor.

In section 6.2.3.1, the average allocation factor $\tilde{c}_{pq}$ was calculated from allocation factors c_{pq}^{κ} referring to different allocation criteria. Here in contrast, an average allocation factor $\tilde{c}_{pq}$ should be calculated from different economic allocation factors c_{pq}^{κ} that correspond with different market prices. These allocation factors can be calculated from historic market prices that represent the fluctuation of the market prices.

The variance of an average economic allocation factor $\tilde{c}_{pq}$ can be calculated from Equations 6.47 using a total number K of economic allocation factors. The corresponding covariance of two average economic allocation factors $\tilde{c}_{pq}$ and $\tilde{c}_{pq'}$ can be calculated from Equation 6.48. Those variances and covariances can be used in Equation 6.26 to propagate the uncertainty due economic allocation with constant allocation factors. Thereby, the contribution of that uncertainty to the overall uncertainty of LCA-results can be quantified.

6.2.3.3 An uncertainty model for allocation between two functional flows

The contribution of uncertainty due to allocation in Equation 6.26 simplifies for allocation of a multi-functional unit process with $\beta = 2$ functional flows. A multi-functional process z with $\beta = 2$ functional flows is divided into two mono-functional processes x and y using two allocation factors c_{zx} and c_{zy}. The condition for allocation factors given in Equation 4.33 defines all K allocation factors c^{κ}_{zy} and the corresponding average allocation factor $\tilde{c}_{pq}$ to

$$c^{\kappa}_{zy} = 1 - c^{\kappa}_{zx} \quad \text{and} \quad \tilde{c}_{zy} = 1 - \tilde{c}_{zx}. \tag{6.49}$$

Following Equation 6.47, the variances of the average allocation factors $\tilde{c}_{zx}$ and $\tilde{c}_{zy}$ are

$$\mathrm{var}(\tilde{c}_{zx}) = \frac{1}{K} \sum_{\kappa=1}^{K} (v^{\kappa}_{zx} - \tilde{v}_{zx})^2 \tag{6.50}$$

and

$$\begin{aligned} \mathrm{var}(\tilde{c}_{zy}) &= \mathrm{var}(1 - \tilde{c}_{zx}) \\ &= \frac{1}{K} \sum_{\kappa=1}^{K} ((1 - c^{\kappa}_{zx}) - (1 - \tilde{c}_{zx}))^2 \\ &= \frac{1}{K} \sum_{\kappa=1}^{K} (\tilde{c}_{zx} - c^{\kappa}_{zx})^2 \\ &= \mathrm{var}(\tilde{c}_{zx}). \end{aligned} \tag{6.51}$$

The covariance of the average allocation factors $\tilde{c}_{zx}$ and $\tilde{c}_{zy}$ is obtained from Equation 6.48, i.e.,

$$\begin{aligned} \mathrm{cov}(\tilde{c}_{zx}, \tilde{c}_{zy}) &= \frac{1}{K}\sum_{\kappa=1}^{K}\left(c_{zx}^{\kappa} - \tilde{c}_{zx}\right)\left(c_{zy}^{\kappa} - \tilde{c}_{zy}\right) \\ &= \frac{1}{K}\sum_{\kappa=1}^{K}\left(c_{zx}^{\kappa} - \tilde{c}_{zx}\right)\left((1 - c_{zx}^{\kappa}) - (1 - \tilde{c}_{zx})\right) \\ &= -\frac{1}{K}\sum_{\kappa=1}^{K}\left(c_{zx}^{\kappa} - \tilde{c}_{zx}\right)^2 \\ &= -\mathrm{var}(\tilde{c}_{zx}). \end{aligned} \tag{6.52}$$

With Equations 6.50, 6.51, and 6.52, the uncertainty contribution of allocation factors in Equation 6.26 simplifies to

$$\begin{aligned} \zeta(h_k, c_{zx}, c_{zy})\mathrm{var}(h_k) &= \sum_{p,q}\left(\frac{\partial h_k}{\partial c_{pq}}\right)^2 \mathrm{var}(c_{pq}) + 2\sum_{p,q}\sum_{q'=q+1}\frac{\partial h_k}{\partial c_{pq}}\frac{\partial h_k}{\partial c_{pq'}}\mathrm{cov}(c_{pq}, c_{pq'}) \\ &= \left(\frac{\partial h_k}{\partial \tilde{c}_{zx}}\right)^2 \mathrm{var}(\tilde{c}_{zx}) + \left(\frac{\partial h_k}{\partial \tilde{c}_{zy}}\right)^2 \mathrm{var}(\tilde{c}_{zy}) \\ &\quad + 2\frac{\partial h_k}{\partial \tilde{c}_{zx}}\frac{\partial h_k}{\partial \tilde{c}_{zy}}\mathrm{cov}(\tilde{c}_{zx}, \tilde{c}_{zy}) \\ &= \left[\left(\frac{\partial h_k}{\partial \tilde{c}_{zx}}\right)^2 + \left(\frac{\partial h_k}{\partial\ \tilde{c}_{zy}}\right)^2 - 2\frac{\partial h_k}{\partial \tilde{c}_{zx}}\frac{\partial h_k}{\partial \tilde{c}_{zy}}\right]\mathrm{var}(\tilde{c}_{zx}) \\ &= \left[\left(\frac{\partial h_k}{\partial \tilde{c}_{zx}}\right) - \left(\frac{\partial h_k}{\partial \tilde{c}_{zy}}\right)\right]^2 \mathrm{var}(\tilde{c}_{zx}). \end{aligned} \tag{6.53}$$

6.3 Discussion and outlook

In this chapter, an analytical approach is presented to assess uncertainties due to fixing multi-functionality problems. Based on the matrix formulation from chapter 4, sensitivity coefficients are derived with respect to aggregation factors (section 6.1.1) and allocation factors (section 6.1.2). These coefficients can be used to perform sensitivity analysis (section 6.1.4) and uncertainty analysis (section 6.2). For uncertainty analysis, a parametric representation of uncertainties due to fixing multi-functionality problems (Table 3.2) is introduced in sections 6.2.2 and 6.2.3. Analytical analysis of these uncertainties is then performed by applying first-order error propagation to Equation 4.42 (section 6.2.1).

The method presented in section 6.2.1 allows quantifying uncertainties in process data, characterization factors, and due to fixing multi-functionality problems. Thus, the analytical approach for uncertainty analysis allows to compare uncertainties from different sources in a holistic method (Equation 6.26). Moreover, the contributions to overall uncertainty in LCA-results can be quantified.

A comparison of the uncertainty contributions from process data, characterization factors, and fixing multi-functionality problems identifies whether LCA-results are dominated by potentially ambiguous choices due to fixing multi-functionality problems. Based on this information, further data refinement is justified if the contribution of process data is dominant. If the contribution from fixing multi-functionality dominates, further data refinement still reduces the overall uncertainty, but the effect can be overshadowed up by uncertainties from fixing multi-functionality.

Still, the choice between system expansion, avoided burden, and allocation is not modeled in Equation 6.26. But the presented approach allows comparing the uncertainty contributions of each method. For this purpose, a multi-functionally problem would be independently fixed with each method. Equation 6.26 would then be examined for uncertainties in LCIA-results from each method: The contributions of uncertainty in aggregation factors used in system expansion or avoided burden would be compared to the contribution of uncertainty in allocation factors. Thereby, it would be determined which method would introduce more uncertainty to LCIA-results.

Even though the formulas for calculating sensitivity coefficients derived in section 6.1 look complicated, the computational effort remains reasonable: all formulas summarized in Table 6.1 require only a single inversion of the fixed technology matrix $\mathbf{A}_{\text{fixed}}$ (cf. Equation 6.6). This matrix inversion is required for calculating LCA-results anyhow (cf. Equations 4.40 to 4.42). In contrast, any approach using MC-simulation requires numerous matrix inversions of the technology matrix. Imbeault-Tétreault et al. (2013) recently compared MC-simulations with analytical propagation of uncertainties in a LCA-study. These authors concluded that the analytical approach is indeed much faster and can be systematically automated in existing software tools. Implementing Equation 6.26 in LCA-software tools could systematize analysis of uncertainties due to fixing multi-functionality problems because the uncertainties can be calculated parallel with the LCA-results, if input uncertainties are specified.

In future work, a comparison similar to that by Imbeault-Tétreault et al. (2013)

should include uncertainties due to fixing multi-functionality problems in order to demonstrate the applicability for large and thus real-world process systems. In real-world process systems with several multi-functional unit processes, the dimensions of the matrices $\mathbf{E}$, $\mathbf{V}$, $\mathbf{T}_1$, $\mathbf{T}_0$, $\mathbf{U}$, and $\mathbf{C}$ will increase leading to higher computational effort, too. Therefore, it seems particularly interesting to investigate how the presented approach performs when several multi-functional unit processes cause several multi-functionality problems within a single LCA.

A drawback of the presented analytical approach is that the first-order approximation in Equation 6.26 considers only uncertainties in the vicinity of the nominal parameter values: The propagation of large uncertainties by Equation 6.26 is thus not exact (Morgan and Henrion (1992) on p. 185). This drawback does not exist for MC-simulations. Large uncertainties can be found, e.g., for the selection between allocation criteria. But the error in the first-order propagation is less severe for smooth mathematical relationships; for linear relationships, first-order approximations are even exact (Morgan and Henrion (1992) on p. 186). LCIA-results depend only linearly on allocation factors calculated from Equation 4.42 as long as no feedback loops for the allocated functional flows exist (cf. Appendix A). In this case, the presented approach for propagating uncertainties in allocation factors is exact and computationally more efficient than MC-simulations.

In case of large uncertainties and non-linear relationships between LCIA-results and allocation factors, higher order approximations could be used for propagating input uncertainties in the allocation factors. Higher order approximations have been applied to the sequential approach (cf. section 3.1.3) for calculating LCA-results; the uncertainties propagated with higher order approximations match those propagated with MC-simulations better than first-order approximations (cf. Ciroth et al. (2004)). Future work could investigate higher order approximations for the matrix approach as well. In such investigations, uncertainties due to fixing multi-functionality problems could be integrated by using the formulations derived in chapter 4.

In Equation 6.26, correlations are included only between aggregation factors and between allocation factors. However, additional covariances exist: Allocation factors often correlate with economic flows that represent the allocation criterion, e.g., mass flows of products. Future research should aim to include these correlations in a mathematical framework for uncertainty calculations in LCA. A starting point could be to express allocation factors (allocation matrix elements c_{pq}) as a function of such economic flows (technology matrix elements a_{ij}).

In a comparative LCA, the LCA-results of competing alternatives share some process data. If the shared process data includes uncertainties due to fixing multi-functionality problems, these uncertainties may cancel each other in the comparative LCA-result: The single LCA-results of each competing alternative are in fact correlated. In future work, the correlation of such comparative LCA-results should be investigated focusing on uncertainties due to fixing multi-functionality problems. Hong et al. (2010) and Imbeault-Tétreault et al. (2013) use comparative metrics as LCA-result that include correlations between shared process data. The approach could be expanded in future work to include correlated uncertainties due to fixing multi-functionality problems.

The approach presented in this chapter is applied in a comparative LCA-study of chlor-alkali electrolysis in the following chapter.

Chapter 7

Comparative LCA of chlor-alkali electrolysis using ODCs

This chapter contains a comparative LCA-study of the two alternative chlor-alkali electrolysis technologies introduced in chapter 2: the membrane process and the ODC-process. These two processes are difficult to compare because they have the non-common product hydrogen that causes multi-functionality problems (section 2.3). Therefore, the procedure for fixing multi-functionality problems developed in chapter 5 is applied. The methods used to fix the multi-functionality problems in this LCA-study introduce uncertainties to the LCA-results. These uncertainties are analyzed using the methods proposed in chapter 6.

The structure of this chapter follows the steps of ISO (2006a) (Figure 3.2): section 7.1 contains goal and scope definition together with LCI-analysis. The goal and scope of this LCA-study is described in section 7.1.1, wherein to the problem statement from section 2.3 is picked up again. In section 7.1.2, three scenarios for alternative functional units are derived based on the alternative distinctions between main products and by-products (section 5.2.1). Subsequently, the system boundaries, inventory data, and impact assessment methods used in this LCA-study are described (section 7.1.3 and 7.1.4). The workflow for fixing multi-functionality problems (Figure 5.5) is applied to each functional unit scenario in section 7.1.5.

In section 7.2, the LCIA-results are presented for each functional unit scenario. Section 7.2.1 contains the results for a functional unit based on the main products chlorine and caustic soda. In section 7.2.2, results are presented for a functional unit based on three main products: chlorine; caustic soda; and hydrogen. The results for a functional unit based on chlorine being the only main product are shown in section 7.2.3.

The LCA-study continues in section 7.3 with a discussion of the LCIA-results of the three scenarios (sections 7.3.1, 7.3.2, and 7.3.3). In those three scenarios, all types of uncertainty due to fixing multi-functionality problems exist (Table 3.2). Therefore, the methods for sensitivity and uncertainty analysis derived in chapter 6 are applied. In section 7.3.1, normalized sensitivity coefficients (section 6.1.4) are calculated and the uncertainty due to a constant mix process as avoided process is examined (section 6.2.2.1. In section 7.3.2, uncertainties due to selection of discrete marginal processes (section 6.2.2.2) and due to additional process data are discussed (section 6.2.2.4). Finally, uncertainties due to allocation (sections 6.2.3.1 and 6.2.3.2) are evaluated in section 7.3.3.

7.1 Goal and scope definition and LCI-analysis

7.1.1 Goal and scope

This LCA-study aims to compare the environmental impacts of two chlor-alkali electrolysis technologies: the membrane process and the ODC-process, cf. chapter 2. The LCIA focuses on the GWI. The results should identify the processes and parameters in the life cycle of the two processes that have considerable influence on this impact category. Other impact categories than global warming should complete the picture of this comparative LCA-study. Conditions and settings should be identified where the ODC-process is environmentally most beneficial.

The cathode is the major difference in the design of the membrane process and the ODC-process, cf. Figures 2.1 and 2.2. In fact, both processes use the same membrane. To identify the true difference between both technologies in the name, the membrane process described in section 2.1 is abbreviated as standard cathode (STC)-process in this chapter.

7.1.2 Functional unit scenarios

Comparative LCA-studies should be based on the same functional unit (ISO (2006b) on p. 22). The STC- and ODC-processes share the products Cl_2 and NaOH but H_2 is a non-common product. It is unclear if the non-common product H_2 should be included in the functional unit. This problem is solved by applying the procedure for comparative LCA of multi-product processes (cf. chapter 5). In section 5.2.1, it is proposed to distinguish main and by-products and to include main products in the functional unit.

The distinction of the products Cl_2, NaOH, and H_2 depends on the site specific settings. The utilization of H_2 varies from site to site, which makes a universally valid distinction of the products difficult. Jörissen et al. (2011) estimate that half of the utilized hydrogen is burned as fuel and half processed as commodity. If recycled as fuel, hydrogen is usually mixed with natural gas and burned for electricity and/or heat generation. As a chemical commodity, hydrogen is mainly used as reagent for synthesis of ammonia and methanol, or for amine production as an intermediate step of polymer production (O'Brien et al., 2005f).

In this work, two alternative scenarios for distinguishing main and by-products are introduced to account for the different hydrogen utilization pathways. In addition, a third scenario refers the CFP of Cl_2. Each scenario corresponds with a different functional unit. The scenarios are introduced in the following paragraphs and summarized in Table 7.1.

In the first scenario, both chlorine and caustic soda are the reason for operation of chlor-alkali electrolysis plants (O'Brien et al., 2005d). Following section 5.2.1, Cl_2 and NaOH are main products for both STC- and ODC-processes. The functional unit is "producing 1 t Cl_2 and 1.128 t NaOH in aqueous solution with $w_{NaOH} = 0.5$". Hydrogen from the STC-process is a by-product which is often used as fuel: H_2 is mixed with natural gas and burned for electricity or heat generation. This scenario is thus called H_2-Fuel.

The second scenario refers to chemical production sites, where hydrogen is used as a chemical commodity instead of being burned. Therefore, hydrogen from a STC-process can also be seen as third main product in addition to Cl_2 and NaOH. A second scenario is thus called H_2-Commoditiy. Its functional unit is "producing 1 t Cl_2 and 1.128 t NaOH in aqueous solution with $w_{NaOH} = 0.5$ and 0.028 t H_2". The ODC-process does not produce H_2, i.e., a non-common main-product exists.

Chlor-alkali electrolyzers are often build at the site of the Cl_2-demand, while NaOH is typically further transported. Therefore, Cl_2 is assumed to be the only reason of operation and thus the only main-product in the third scenario. The functional unit is thus "producing 1 t Cl_2". The scenario corresponds to the CFP of Cl_2. The CFP refers to the GWI of a product throughout its life cycle. The third scenario is thus called Cl_2-CFP. In this scenario, only the GWI is calculated and discussed.

Table 7.1: Scenarios and corresponding functional units for the comparative LCA-study of STC- and ODC-electrolysis

scenario	reference flow of functional unit
H_2-Fuel	1 t Cl_2 (gaseous, purity and conditions of STC) + 1.128 t NaOH (solved in H_2O, $w_{NaOH} = 0.5$, conditions of STC)
H_2-Commodity	1 t Cl_2 (gaseous, purity and conditions of STC) + 1.128 t NaOH (solved in H_2O, $w_{NaOH} = 0.5$, conditions of STC) + 0.028 t H_2 (gaseous, purity and conditions of STC)
Cl_2-CFP	1 t Cl_2 (gaseous, purity and conditions of STC)

7.1.3 System boundaries and inventory data

Figure 7.1 shows a process flow diagram for the STC-process system. The diagram is split in three life stages manufacturing, operation, and end of life. In the first stage, manufacturing of the STC-electrolyzer is considered. During operation of the STC-electrolysis, the supply of electricity, NaCl, and H_2O is modeled.

The caustic soda from the STC-process is typically concentrated from a mass fraction of NaOH $\omega_{NaOH} = 32\,\%$ to the market standard of $\omega_{NaOH} = 50\,\%$. Therefore, a concentration of NaOH is also included as well as the steam supply for the concentration process. The vapor produced by the concentration is considered as waste that is blown off in this work because it is unknown if and how this vapor is reused in industrial practice.

A candidate for an avoided process is shown in dashed lines (combustion natural gas (NG)). This process is only required for the scenario H_2-Fuel, it is further explained in section 7.1.5.1. For the end of life stage, the disposal of the electrolyzer is considered.

Figure 7.2 illustrates a process flow diagram for the ODC-process system, it is also split in the life stages manufacturing, operation and end of life. The ODC-process system also includes manufacturing and disposal of the electrolyzer. In contrast to the manufacturing of the STC-electrolyzer, parts of the cathode are recycled into the manufacturing process. For operation, the ODC-process system includes the supply of electricity, NaCl, H_2O, and the concentration of NaOH as described for STC-process system.

In addition to STC-process system, the supply of O_2 from an air-fractionation process is included. The air-fractionation process has a second product, i.e., N_2.

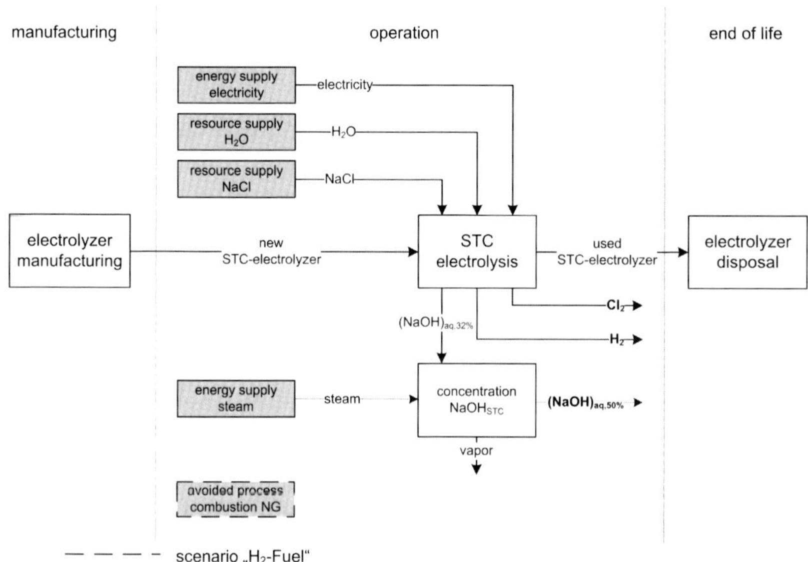

Figure 7.1: Process flow diagram of the STC-process system. Processes displayed in white boxes are unit processes modeled in this work. Processes displayed in grey boxes are taken from on available life cycle inventory data (Table 7.1). The avoided process combustion NG (dashed box) is only used in the scenario H_2-Fuel.

Moreover, two candidate processes for hydrogen production are shown in dot-dashed and dotted lines, respectively: a steam reforming process and a H_2O-electrolysis. These processes is only required in the scenario H_2-Commodity and therefore further explained in section 7.3.2. The H_2O-electrolysis can also supply the stoichiometrically required amount of O_2 for the ODC-process. Therefore, the air-fractionation is dropped if the H_2O-electrolysis is used.

Figure 7.3 shows a process flow diagram for the electrolyzer manufacturing. The electrolyzer consists of the electrolysis cells, a power bus, pipes, and a steel frame. The resource required for these parts are included in the manufacturing process. The major difference between STC- and ODC-electrolyzer is the cathode compartment of the electrolysis cell because STC and ODC are made from different materials. The manufacturing process of ODCs requires noble metals such as silver while STCs are

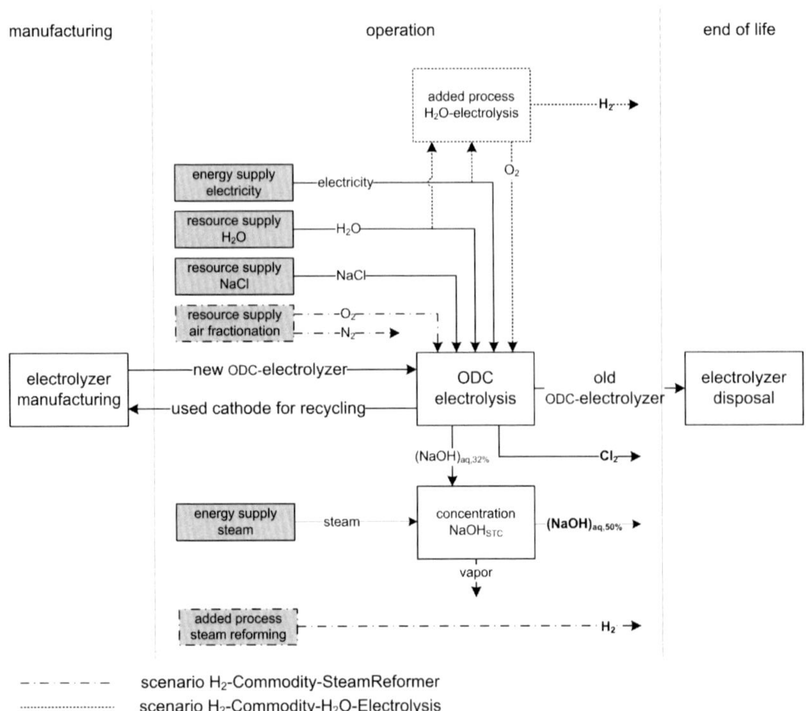

Figure 7.2: Process flow diagram of the ODC-process system. Processes displayed in white boxes are unit processes modeled in this work. Processes displayed in grey boxes are taken from on available life cycle inventory data (Table 7.1). The added processes steam reforming and H_2O-electrolysis (dot-dashed boxes) are only used in the scenario H_2-Commodity.

made from nickel.

For the disposal of the electrolyzer, only limited data is available: In this work, only disposal of electrolyzer parts from steel, polypropylene (PP), and cupper is included.

The processes displayed as grey boxes in Figures 7.1, 7.2, and 7.3 are modeled based on available life cycle inventory data. Table 7.2 shows these life cycle inventory data and their sources. Life cycle inventory data of a unit process consists of elementary flows scaled to the functional flow(s) of the unit process. The functional flow is

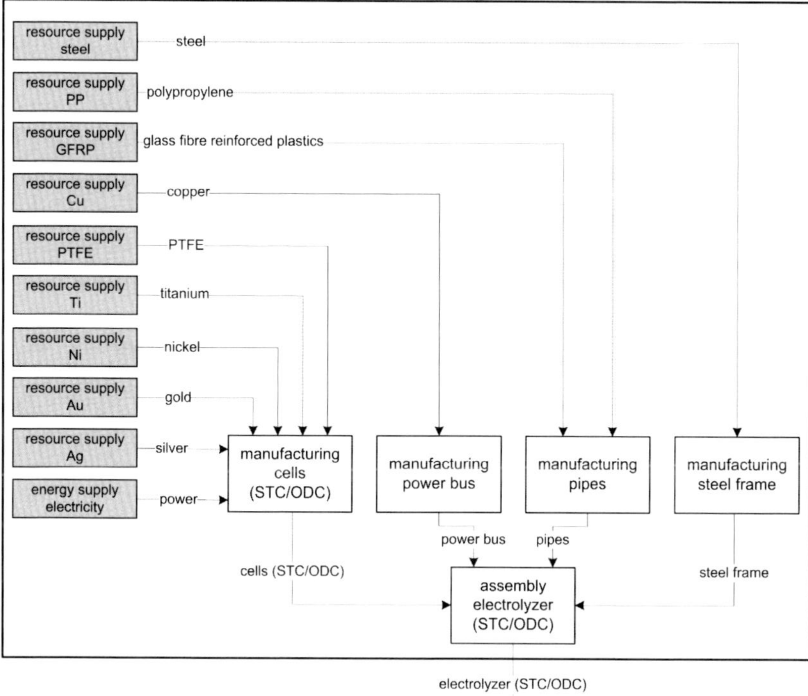

Figure 7.3: Process flow diagram of manufacturing of STC- and ODC-electrolyzer. Processes displayed in white boxes are unit processes modeled in this work. Processes displayed in grey boxes are taken from on available life cycle inventory data (Table 7.1).

typically given as production of one default unit.

The processes displayed as white boxes in Figures 7.1, 7.2, and 7.3 are unit processes that are modeled in this work. The models are based on literature data, physical relationships, and data collections as described in the following paragraphs.

The STC- and ODC-processes are modeled using Equations 2.1, 2.12, and 2.8. The concentration of NaOH is modeled following Sattler (2001). The electricity demand of the H_2O-electrolysis (Figure 7.2) is taken from Table 1 in Ursua et al. (2012). The materials required for manufacturing the electrolyzer components and the recycling process for ODCs was modeled based on primary data from Bayer Material Science (2012).

The processes modeled in this work include only economic flows because elementary flows were not available for these unit processes.

Both STC- and ODC-electrolyzer have an assumed lifetime of 24 years. Therefore, this LCA-study analyzes one lifetime of both STC- and ODC-processes. Both electrolyzers consist of 160 cells. The cathodes and the membranes are assumed to have a lifetime of 4 years, i.e. they are manufactured 6 times within the electrolyzer lifetime. The anode has a lifetime of 8 years and is manufactured three times during an electrolyzer lifetime.

The economic flows of all processes shown in Figures 7.1 and 7.2 are summarized in the technology matrix $\mathbf{A}_{\text{CA}}$ shown in Table 7.3. The index of the matrix $\mathbf{A}_{\text{CA}}$ refers to chlor-alkali (CA)-process system. In the technology matrix $\mathbf{A}_{\text{CA}}$, the unit processes based on life cycle inventory data (Table 7.2) have only one output, i.e., their functional flow. In contrast, the unit processes modeled in this work contain input and output flows. In the matrix $\mathbf{A}_{\text{CA}}$ the manufacturing and disposal of each electrolyzer is summarized in a single unit process named manufacturing.

The three functional units defined in the previous section (cf. Table 7.1) correspond with three final demand matrices. These final demand matrices are shown in Table 7.4. The final demand matrix $\mathbf{F}_{\text{H}_2\text{-Fuel}}$ corresponds with the scenario H_2-Fuel; the matrix contains the main product flows Cl_2 and NaOH of the STC- and ODC-process.

The final demand matrix $\mathbf{F}_{\text{H}_2\text{-Commodity}}$ relates to the scenario H_2-Commodity and includes H_2 as a third main product flow from the STC-process. The parameters φ_1 and φ_2 indicate that H_2-flows of the steam reforming process and the H_2O-electrolysis could be used for system expansion. This is further discussed in section 7.1.5.2.

The final demand matrix $\mathbf{F}_{\text{Cl}_2\text{-CFP}}$ refers to the scenario Cl_2-CFP and contains only the main product flow Cl_2 from both STC- and ODC-processes.

Table 7.2: Life cycle inventory data used for modeling unit processes in the chlor-alkali process system (unit processes displayed as grey boxes in Figures 7.1 to 7.3.

process	data set	data source
resource supply O_2	oxygen (DE)	GaBi-PE (2012)
resource supply H_2O	water, completely softened, at plant (RER)	Ecoinvent (2008)
resource supply NaCl	sodium chloride, powder, at plant (RER)	Ecoinvent (2008)
resource supply steam	steam, for chemical processes, at plant (RER)	Ecoinvent (2008)
energy supply electricity	electricity, medium voltage, at grid (DE)	Ecoinvent (2011)
resource supply steel	steel, low-alloyed, at plant (RER)	Ecoinvent (2008)
resource supply PP	polypropylene, granulate, at plant (RER)	Ecoinvent (2008)
resource supply GFRP	glass fibre reinforced plastic, polyester resin, hand lay-up, at plant (RER)	Ecoinvent (2011)
resource supply PTFE	tetrafluoroethylene, at plant (RER)	Ecoinvent (2008)
resource supply Cu	copper, primary, at refinery (RER)	Ecoinvent (2011)
resource supply Ti	titanium (GLO)	GaBi-PE (2012)
resource supply Ni	nickel, 99.5%, at plant (GLO)	Ecoinvent (2008)
resource supply Ag	silver, at regional storage (RER)	Ecoinvent (2008)
resource supply Au	gold, at regional storage (RER)	Ecoinvent (2011)
steam reforming	hydrogen (DE)	GaBi-PE (2012)
combustion NG	natural gas, burned in power plant (DE)	Ecoinvent (2008)
disposal steel parts	disposal, steel, 0% water, to municipal incineration (CH)	Ecoinvent (2008)
disposal power bus	disposal, copper, 0% water, to municipal incineration (CH)	Ecoinvent (2008)
disposal pipes	disposal, polypropylene, 15.9% water, to municipal incineration)	Ecoinvent (2008)

Table 7.3: Technology matrix $\mathbf{A}_{CA}$ of the chlor-alkali process system

	unit	manufacturing STC	manufacturing ODC	energy supply electricity	resource supply H_2O	resource supply NaCl	air-fractionation	energy supply steam	STC electrolysis	concentration NaOH, STC	ODC electrolysis	concentration NaOH, ODC	steam reforming	H_2O-electrolysis	combustion NG (mix GER)
STC electrolyzer	-	1	0	0	0	0	0	0	-1	0	0	0	0	0	0
ODC electrolyzer	-	0	1	0	0	0	0	0	0	0	-1	0	0	0	0
electricity	kWh	0	0	1	0	0	0	0	-2338	0	-1558	0	0	-47831	0
H_2O	t	0	0	0	1	0	0	0	-0.508	0	-0.254	0	0	-8.94	0
NaCl	t	0	0	0	0	1	0	0	-1.648	0	-1.648	0	0	0	0
O_2	t	0	0	0	0	0	1	0	0	0	-0.226	0	0	0	0
N_2	t	0	0	0	0	0	3.255	0	0	0	0	0	0	0	0
steam	t	0	0	0	0	0	0	1	0	-1.125	0	-1.125	0	0	0
$Cl_{2,STC}$	t	0	0	0	0	0	0	0	1	0	0	0	0	0	0
$NaOH_{STC,32}$	t	0	0	0	0	0	0	0	1.128	-1	0	0	0	0	0
$NaOH_{STC,50}$	t	0	0	0	0	0	0	0	0	1	0	0	0	0	0
$H_{2,STC}$	t	0	0	0	0	0	0	0	0.028	0	0	0	0	0	0
$Cl_{2,ODC}$	t	0	0	0	0	0	0	0	0	0	1	0	0	0	0
$NaOH_{ODC,32}$	t	0	0	0	0	0	0	0	0	0	1.128	-1	0	0	0
$NaOH_{ODC,50}$	t	0	0	0	0	0	0	0	0	0	0	1	0	0	0
$H_{2,SR}$	t	0	0	0	0	0	0	0	0	0	0	0	1	0	0
$H_{2,EL}$	t	0	0	0	0	0	0	0	0	0	0	0	0	1	0
$O_{2,EL}$	t	0	0	0	0	0	0	0	0	0	0	0	0	7.94	0
Natural gas (mix GER)	t H_2-eq.	0	0	0	0	0	0	0	0	0	0	0	0	0	1

Table 7.4: Final demand matrices $\mathbf{F}_{CA}$ for scenarios H_2-Fuel, H_2-Commodity, and Cl_2-CFP of the chlor-alkali process system

economic flows	unit	$\mathbf{F}_{H_2\text{-Fuel}}$		$\mathbf{F}_{H_2\text{-Commodity}}$		$\mathbf{F}_{Cl_2\text{-CFP}}$	
		f_{STC}	f_{ODC}	f_{STC}	f_{ODC}	f_{STC}	f_{ODC}
STC electrolysis plant	-	0	0	0	0	0	0
ODC electrolysis plant	-	0	0	0	0	0	0
electricity	kWh	0	0	0	0	0	0
H_2O	t	0	0	0	0	0	0
NaCl	t	0	0	0	0	0	0
O_2	t	0	0	0	0	0	0
N_2	t	0	0	0	0	0	0
steam	t	0	0	0	0	0	0
$Cl_{2,STC}$	t	1	0	1	0	1	0
$NaOH_{STC,32}$	t	0	0	0	0	0	0
$NaOH_{STC,50}$	t	1.128	0	1.128	0	0	0
$H_{2,STC}$	t	0	0	0.028	0	0	0
$Cl_{2,ODC}$	t	0	1	0	1	0	1
$NaOH_{ODC,32}$	t	0	0	0	0	0	0
$NaOH_{ODC,50}$	t	0	1.128	0	1.128	0	0
$H_{2,SR}$	t	0	0	0	φ_1	0	0
$H_{2,EL}$	t	0	0	0	φ_2	0	0
$O_{2,EL}$	t	0	0	0	0	0	0
Natural gas (mix GER)	t H_2-eq.	0	0	0	0	0	0

7.1.4 Impact assessment method

In this work, seven environmental impact categories are assessed:

- global warming (GW) in CO_2-equivalent emissions,
- ozone depletion in CFC-11-equivalent emissions,
- acidification in SO_2-equivalent emissions,
- eutrophication in P-equivalent emissions,
- photochemical ozone creation in NMVOCemissions,
- human toxicity in 1,4 DB-equivalent emissions,
- fossil resource depletion in oil-equivalents.

These seven impact categories were selected during a meeting with researchers and stakeholders developing the ODC-technology (Bayer Material Science, 2010b). The impact categories are modeled using the impact assessment method ReCiPe (Goedkoop et al., 2013). The ReCiPe is the follow up of two well-known LCIA-methods, i.e., Eco-indicator 99 and CML 2002; its impact categories are described in reviewed magazines (European Commission (2010a) on pp. 40-43).

For a better illustration, the elementary flows of the unit processes (cf. Figures 7.1 to 7.3) are directly multiplied by the characterization factors of the seven impact categories. In matrix-based LCA, this multiplication is written as a matrix product of an intervention matrix $\mathbf{B}$ and a characterization matrix $\mathbf{Q}$ (section 3.1.3). Therefore, the result of the product is referred to as matrix $(\mathbf{QB})_{\mathrm{CA}}$ shown in Table 7.5.

The matrix $(\mathbf{QB})_{\mathrm{CA}}$ contains the environmental impact flows qb_{kj} for the seven environmental impact categories mentioned above. However, only the processes modeled with life cycle inventory data (displayed as grey boxes in Figures 7.1 to 7.3, cf. Table 7.2) contain elementary flows. Therefore, only these processes contain environmental impact flows qb_{kj}. The unit processes modeled in this work (displayed as grey boxes in Figures 7.1 to 7.3) do not include elementary flows, the corresponding environmental impact flows are thus $qb_{kj} = 0$ (Table 7.5).

7.1.5 Fixing multi-functionality problems

In this section, multi-functionality problems are fixed for each scenario. Fixing multi-functionality problems follows the workflow derived in chapter 5 (Figure 5.5).

7.1.5.1 Scenario H_2-Fuel

In the scenario H_2-Fuel, the main products are Cl_2 and NaOH for both STC- and ODC-process (section 7.1.2). The corresponding main product technology matrix $\mathbf{A}_{\mathrm{CA}}^{\mathrm{mp,H_2\text{-}Fuel}}$ contains only the economic flows of Cl_2 and NaOH from the STC- and ODC-process. Therefore, the first part of the workflow (Figure 5.2) is concluded because the main product discrepancy matrix becomes

$$\mathbf{D}^{\mathrm{mp,H_2\text{-}Fuel}} = \mathbf{A}_{\mathrm{CA}}^{\mathrm{mp,H_2\text{-}Fuel}} \cdot \left(\mathbf{A}_{\mathrm{CA}}^{\mathrm{mp,H_2\text{-}Fuel}}\right)^{+} \cdot \mathbf{F}_{\mathrm{H_2-Fuel}} - \mathbf{F}_{\mathrm{H_2-Fuel}} = \mathbf{0}. \tag{7.1}$$

The second part of the workflow (Figure 5.4) continues with calculation of the discrepancy matrix. The discrepancy matrix $\mathbf{D}^{\mathrm{H_2\text{-}Fuel}}$ is shown in Table 7.6. It reveals two non-common by-products because

Table 7.5: Intervention and characterization matrices summarized as matrix $(\mathbf{QB})_{CA}$ for the chlor-alkali process system

	unit	manufacturing STC	manufacturing ODC	energy supply electricity	resource supply H_2O	resource supply NaCl	air-fractionation	energy supply steam	STC electrolysis	concentration NaOH, STC	ODC electrolysis	concentration NaOH, ODC	steam reforming	H_2O-electrolysis	combustion NG (mix GER)
CO_2-equiv.	t	2.00E-03	2.47E-03	6.80E-04	2.47E-05	2.03E-01	7.06E-01	2.77E-01	0	0	0	0	1.11E+01	0	8.20E+00
CFC-11-equiv.	g	2.65E-02	3.66E-02	2.86E-05	2.33E-06	1.24E-02	3.46E-03	0	0	0	0	0	5.97E-04	0	1.33E+00
SO_2-equiv.	kg	5.38E-02	2.66E-02	8.36E-04	6.45E-05	8.82E-01	1.01E+00	1.99E+00	0	0	0	0	6.49E+00	0	8.49E+00
P-equiv.	kg	5.14E-03	5.20E-03	8.42E-04	7.36E-07	1.48E-02	5.31E-05	1.97E-06	0	0	0	0	8.92E-04	0	5.05E-02
NMVOC-equiv.	kg	1.16E-02	9.33E-03	7.05E-04	7.01E-05	5.25E-01	8.29E-01	8.30E-01	0	0	0	0	7.09E+00	0	1.16E+01
1,4-DB-equiv.	t	2.84E-03	2.78E-03	9.31E-05	2.15E-06	9.71E-03	3.49E-03	1.28E-04	0	0	0	0	1.24E-03	0	5.84E-02
oil-equiv.	t	3.55E-04	3.86E-04	1.84E-04	6.21E-06	7.01E-02	1.83E-01	8.92E-02	0	0	0	0	5.19E+00	0	3.53E+00

$$\left(d_{\text{H}_2\text{-Fuel}}\right)_{\text{H}_2,\text{STC}} > 0 \text{ and } \left(d_{\text{H}_2\text{-Fuel}}\right)_{\text{N}_2,\text{ODC}} > 0. \tag{7.2}$$

The non-common by-products are H_2 from the STC-process and N_2 from the air-fractionation process. Following the workflow the site-specific use of the non-common by-products has to be identified before avoided burden is preferably applied (Figure 5.4).

For H_2 from the STC-process, the site-specific use is already described in section 7.1.2: H_2 typically avoids combustion of natural gas. Thus, combustion of natural gas is applied as avoided process. The avoided process is displayed as a dashed box in Figure 7.1. It is modeled with the life cycle inventory data for the combustion of the German natural gas mix (Table 7.2). That dataset uses a constant mix composition for the origin of the natural gas within the German mix. The uncertainty due to using this constant mix composition is analyzed in section 7.3.1.

The by-product N_2 is typically used as a chemical commodity at chemical production sites (Bayer Material Science, 2010a). Following the procedure in Figure 5.5, avoided burden should be applied. However, no avoided process for is available N_2 in this work. Therefore, allocation is applied to fix the multi-functionality problem. Here, the mass fraction of N_2 and O_2 was used as allocation criterion because market prices for economic allocation were not available. Moreover, mass-based allocation is in line with the life cycle inventory data used to model the air-fractionation process (Table 7.2).

Avoided burden for H_2 and allocation for N_2 can be applied using the matrix formulation presented in chapter 4. A recalculation of the discrepancy matrix yields a discrepancy matrix $\mathbf{D}_{\text{H}_2\text{-Fuel}} = \mathbf{0}$ and concludes fixing the multi-functionality problem of the scenario H_2-Fuel.

7.1.5.2 Scenario H_2-Commodity

In the scenario H_2-Commodity, the main products are Cl_2, NaOH, and H_2 (section 7.1.2). The STC-process produces all three main products, but the ODC-process does not: H_2 is a non-common main product. The corresponding main product discrepancy matrix is

$$\begin{aligned} \mathbf{D}^{\text{mp},\text{H}_2\text{-Commodity}} &= \mathbf{A}_{\text{CA}}^{\text{mp},\text{H}_2\text{-Commodity}} \cdot \left(\mathbf{A}_{\text{CA}}^{\text{mp},\text{H}_2\text{-Commodity}}\right)^{+} \cdot \mathbf{F}_{\text{H}_2\text{-Commodity}} \\ &\quad -\mathbf{F}_{\text{H}_2\text{-Commodity}} \neq \mathbf{0}, \end{aligned} \tag{7.3}$$

Table 7.6: Discrepancy matrices $\mathbf{D}_{CA}$ for scenarios H_2-Fuel, H_2-Commodity, and Cl_2-CFP of the chlor-alkali process system

economic flows	unit	$\mathbf{D}_{H_2\text{-Fuel}}$		$\mathbf{D}_{H_2\text{-Commodity}}$		$\mathbf{D}_{Cl_2\text{-CFP}}$	
		$\mathbf{f}_{STC}$	$\mathbf{f}_{ODC}$	$\mathbf{f}_{STC}$	$\mathbf{f}_{ODC}$	$\mathbf{f}_{STC}$	$\mathbf{f}_{ODC}$
STC electrolysis plant	-	0	0	0	0	0	0
ODC electrolysis plant	-	0	0	0	0	0	0
electricity	kWh	0	0	0	0	0	0
H_2O	t	0	0	0	0	0	0
NaCl	t	0	0	0	0	0	0
O_2	t	0	-0.2	0	-0.2	0	-0.12
N_2	t	0	0.06	0	0.06	0	0.04
steam	t	0	0	0	0	0	0
$Cl_{2,STC}$	t	-5E-4	0	0	0	-0.4	1
$NaOH_{STC,32}$	t	-3E-4	0	0	0	0.34	0
$NaOH_{STC,50}$	t	-3E-4	0	0	0	0.34	0
$H_{2,STC}$	t	0.03	0	0	0	0.02	0
$Cl_{2,ODC}$	t	0	-0.03	0	-0.03	0	-0.410
$NaOH_{ODC,32}$	t	0	-0.02	0	-0.02	0	0.34
$NaOH_{ODC,50}$	t	0	-0.02	0	-0.02	0	0.34
$H_{2,SR}$	t	0	0	0	0	0	0
$H_{2,EL}$	t	0	0	0	0	0	0
$O_{2,EL}$	t	0	0	0	0	0	0
Natural gas (mix GER)	t H_2-eq.	0	0	0	0	0	0

the multi-functionality problem due to non-common main products. System expansion is thus required to fix the multi-functionality problem (Figure 5.2).

Two unit processes for H_2-production are considered as candidates for the marginal process: a steam reforming process and a H_2O-electrolysis. Steam reforming is available on an industrial scale and the most common technology for hydrogen production at chemical production sites. H_2O-electrolysis are also available (Ursua et al., 2012), even though they are not common on an industrial scale yet. However, they are considered as candidate for a marginal process in this work because they can co-produce the stoichiometrically required amount of O_2 for the ODC-process. Thereby, the air-fractionation is not required for O_2-supply in combination with a H_2O-electrolysis.

System expansion changes the final demand matrix $\mathbf{F}_{H_2\text{-Commodity}}$ by specifying the parameters φ_1 and φ_2 (Table 7.4). In this work, LCIA-results are calculated for both candidate processes individually. Therefore, the final demand matrix $\mathbf{F}_{H_2\text{-Commodity,exp}}$ consists of three final demand vectors after system expansion with the steam reforming process and the H_2O-electrolysis. This final demand matrix $\mathbf{F}_{H_2\text{-Commodity,exp}}$ is shown in Table 7.7. In section 7.3.2, it is discussed how the uncertainty due to the selection between the steam reforming process and the H_2O-electrolysis can be quantified using the method proposed in sections 6.2.2.2 and 6.2.2.3.

Recalculating the main product discrepancy matrix with the final demand matrix $\mathbf{F}_{H_2\text{-Commodity,exp}}$ yields

$$\begin{aligned}\mathbf{D}^{\text{mp,H}_2\text{-Commodity}} &= \mathbf{A}_{\text{CA}}^{\text{mp,H}_2\text{-Commodity}} \cdot \left(\mathbf{A}_{\text{CA}}^{\text{mp,H}_2\text{-Commodity}}\right)^{+} \cdot \mathbf{F}_{\text{H}_2\text{-Commodity,exp}} \\ &\quad -\mathbf{F}_{\text{H}_2\text{-Commodity,exp}} = \mathbf{0}\end{aligned} \tag{7.4}$$

and concludes the first part of the workflow (Figure 5.2).

Following the second part of the workflow (Figure 5.4), the discrepancy matrix $\mathbf{D}^{H_2\text{-Fuel}}$ is calculated with the expanded final demand matrix $\mathbf{F}_{H_2\text{-Commodity,exp}}$. The discrepancy matrix $\mathbf{D}^{H_2\text{-Fuel}}$ for system expansion with the steam reforming process is shown in Table 7.6. In this scenario, only one non-common by-product remains because only $\mathbf{d}_{N_{2ODC}} > 0$: N_2 from the air-fractionation. Allocation using mass fraction of products as allocation criterion is again used to fix this multi-functionality problem (section 7.1.5.1).

The non-common by-product N_2 occurs solely for system expansion with the steam reforming process. For system expansion with the H_2O-electrolysis, O_2 is fully supplied by the H_2O-electrolysis and the air-fractionation process is not required (not shown in Table 7.6).

A recalculation of the discrepancy matrix yields a discrepancy matrix $\mathbf{D}_{H_2\text{-Fuel}} = \mathbf{0}$ and concludes fixing the multi-functionality problem of this scenario.

7.1.5.3 Scenario Cl_2-CFP

In the scenario Cl_2-CFP, the only main product is Cl_2 (section 7.1.2). Following the workflow (Figure 5.2), the corresponding main product discrepancy matrix $\mathbf{D}^{\text{mp,Cl}_2\text{-CFP}}$ is

$$\mathbf{D}^{\text{mp,Cl}_2\text{-CFP}} = \mathbf{A}_{\text{CA}}^{\text{mp,Cl}_2\text{-CFP}} \cdot \left(\mathbf{A}_{\text{CA}}^{\text{mp,Cl}_2\text{-CFP}}\right)^{+} \cdot \mathbf{F}_{\text{Cl}_2\text{-CFP}} - \mathbf{F}_{\text{Cl}_2\text{-CFP}} = \mathbf{0} \tag{7.5}$$

Table 7.7: Final demand matrix $\mathbf{F}_{H_2\text{-Commodity}}$ for scenario H_2-Commodity after system expansion

economic flows	unit	$\mathbf{F}_{H_2\text{-Commodity}}$		
		f_{STC}	$f_{ODC,SR}$	$f_{ODC,EL}$
STC electrolysis plant	-	0	0	0
ODC electrolysis plant	-	0	0	0
electricity	kWh	0	0	0
H_2O	t	0	0	0
NaCl	t	0	0	0
O_2	t	0	0	0
N_2	t	0	0	0
steam	t	0	0	0
$Cl_{2,STC}$	t	1	0	0
$NaOH_{STC,32}$	t	0	0	0
$NaOH_{STC,50}$	t	1.128	0	0
$H_{2,STC}$	t	0.028	0	0
$Cl_{2,ODC}$	t	0	1	1
$NaOH_{ODC,32}$	t	0	0	0
$NaOH_{ODC,50}$	t	0	1.128	1.128
$H_{2,SR}$	t	0	0.028	0
$H_{2,EL}$	t	0	0	0.028
$O_{2,EL}$	t	0	0	0
Natural gas (mix GER)	t H_2-eq.	0	0	0

because Cl_2 is a common main product of both STC- and ODC-process. The procedure thus continues with calculation of the discrepancy matrix $\mathbf{D}^{Cl_2\text{-CFP}}$ (Figure 5.4). The discrepancy matrix $\mathbf{D}^{Cl_2\text{-CFP}}$ is shown in Table 7.6.

Here, a common by-product NaOH and a non-common by-product H_2 exists. Avoided burden should be used to fix the multi-functionality problem (Figure 5.5). But an avoided process producing NaOH is not available. Therefore, allocation is applied to fix the multi-functionality problem due to both the common by-product NaOH and the non-common by-product H_2.

In this work, four allocation criteria were considered: mass fraction, mole fraction, exergy fraction, and market value fraction of the electrolysis products. The resulting allocation factors are summarized in Table 7.8. At first, LCIA-results are calculated for all four allocation criteria. In the discussion in section 7.3.3, the uncertainty due to selecting between these allocation criteria is analyzed following the method presented in section 6.2.3.1.

Table 7.8: Allocation factors for the STC- and ODC-process products using allocation criteria mass fraction, mole fraction, exergy fraction, and market value fraction

criteria	STC			ODC	
	Cl_2	NaOH	H_2	Cl_2	NaOH
mass fraction	0.464	0.523	0.013	0.470	0.530
mole fraction	0.250	0.500	0.250	0.333	0.667
exergy fraction	0.248	0.285	0.467	0.465	0.535
market value fraction	0.266	0.669	0.065	0.284	0.716

7.2 Impact assessment results

7.2.1 Scenario H_2-Fuel

Figure 7.4 shows the GWI of STC- and ODC-process systems for the scenario H_2-Fuel. For the STC-system, the avoided impacts from combustion of natural gas are shown as negative bar. All other contributions of the STC-system are offset by the impact of the avoided process to illustrate the difference between STC- and ODC-system graphically.

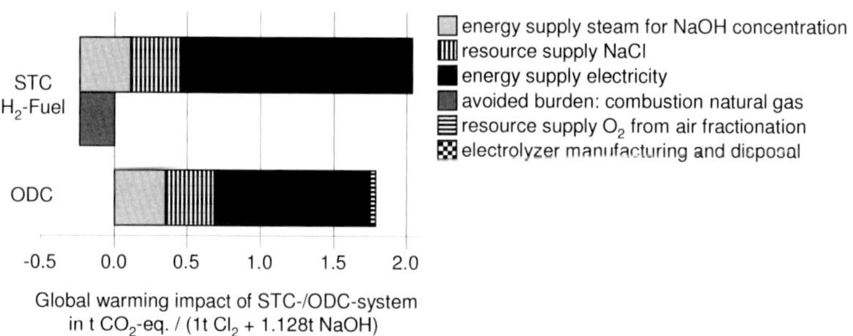

Figure 7.4: GWI for scenario H_2-Fuel applying avoided burden with the avoided process combustion NG, cf. Figure 7.1

The STC-system has an impact of 2.04 t $CO_{2_{eq}}$ emissions per functional unit. The impact of the ODC-system is 12 % lower at 1.80 t $CO_{2_{eq}}$ emissions per functional unit.

The largest contribution to the GWI origins from electricity supply for the STC- and ODC-process (70 % and 59 %). Resource supply of NaCl and steam supply for the concentration process contribute considerably to the GWI, but both processes are equal for STC- and ODC-process. The air-fractionation process for O_2 supply contributes only 2 % to the impact of the ODC-process, while the avoided process reduces the impact of the STC-system by 10 %. The manufacturing and disposal of the electrolyzers contribute only 0.09 % and 0.13 % to the STC- and ODC-system making their bars nearly invisible in Figure 7.4. H_2O-supply is not displayed because its contribution is even smaller.

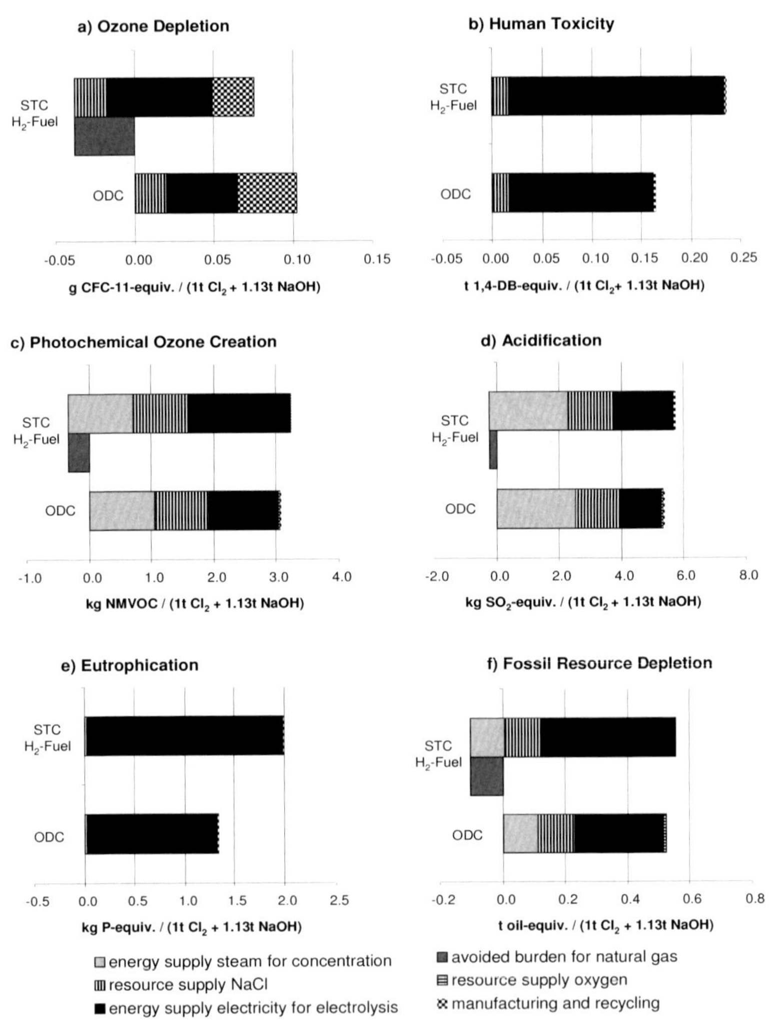

Figure 7.5: Impacts in ozone depletion, human toxicity, photochemical ozone creation, acidification, eutrophication, and fossil resource depletion for scenario H_2-Fuel applying avoided burden with the avoided process combustion NG, cf. Figure 7.1

Figure 7.5 shows the impacts of the STC- and the ODC-systems in six other environmental impact categories: a) ozone depletion, b) human toxicity, c) photochemical ozone creation, d) acidification, e) eutrophication, and f) fossil resource depletion.

The ODC-system has lower impacts in all categories except for ozone depletion potential (ODP): In that category, the ODC-system causes 33 % higher CFC-$11_{eq.}$-emissions than the STC-system. The reason is twofold: First, the avoided process causes a higher offset than in all other considered impact categories. The high ODP-impact from burning natural gas origins from pipeline leakages modeled in the aggregated dataset used in this work (cf. Table 7.2). A second reason for the advantage of the STC-system over the ODC-system is the contribution of electrolyzer manufacturing. Manufacturing of the ODCs requires more polytetrafluoroethylene (PTFE) which causes the higher impacts of manufacturing an ODC-electrolyzer compared to a STC-electrolyzer in this category.

The differences between the STC- and ODC-system in photochemical ozone creation potential (POCP), acidification potential (AP), and fossil resource depletion are smaller than in the GWI: Still, the ODC-system has 6 %, 7 %, and 6 % lower impacts than the STC-system in those categories.

The ODC-process has a large advantage in human toxicity potential (HTP) and eutrophication potential (EP): the impacts of the ODC-system are 43 % and 33 % lower than the impacts of the STC-system. In both categories, the electricity supply for the electrolysis dominates by contributing 88 % and 97 % to the results of the ODC-system.

The environmental impacts of the avoided process influence the overall comparison between STC- and ODC-system (Figures 7.4 and 7.5). However, these environmental impacts of the avoided process are also subject to uncertainties due to the constant mix composition of the avoided process. Therefore, the uncertainty due to using an avoided process with a constant mix process is analyzed using the method from section 6.2.2.1 in section 7.3.1.

7.2.2 Scenario H_2-Commodity

Figure 7.6 shows the GWI of STC- and ODC-process systems for the scenario H_2-Commodity.

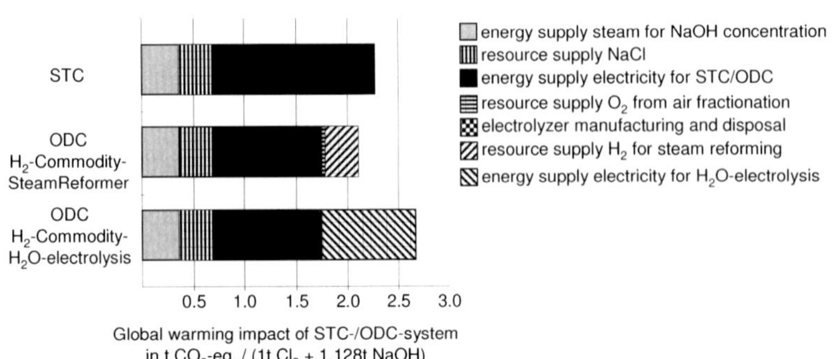

Figure 7.6: GWI for scenario H_2-Commodity applying system expansion with added process steam reforming (ODC H_2-Commodity-SteamReformer) and added process H_2O-electrolysis (ODC H_2-Commodity-H_2O-electrolysis), cf. Figure 7.2

In this scenario, the STC-system has an impact of 2.28 t CO_{2eq} emissions per functional unit. The ODC system has an 8 % lower GWI for system expansion with the steam reforming process (2.1 t CO_{2eq}) but a 17 % higher impact for system expansion with the H_2O-electrolysis (2.67 t CO_{2eq}). The contribution of single processes to the GWI are similar to the scenario H_2-Fuel: the electricity supply for the electrolysis has the highest contribution.

Figure 7.7 shows the impacts of the STC- and the ODC-system in six more environmental impact categories: a) ozone depletion, b) human toxicity, c) photochemical ozone creation, d) acidification, e) eutrophication, and f) fossil resource depletion.

The ODC-system with a steam reforming process has lower impacts than the STC-system in all categories except for fossil resource depletion: In this category, the STC-system has 2 % lower impacts than the ODC-system. In contrast, the ODC-system with H_2O-electrolysis has higher impacts in all six categories.

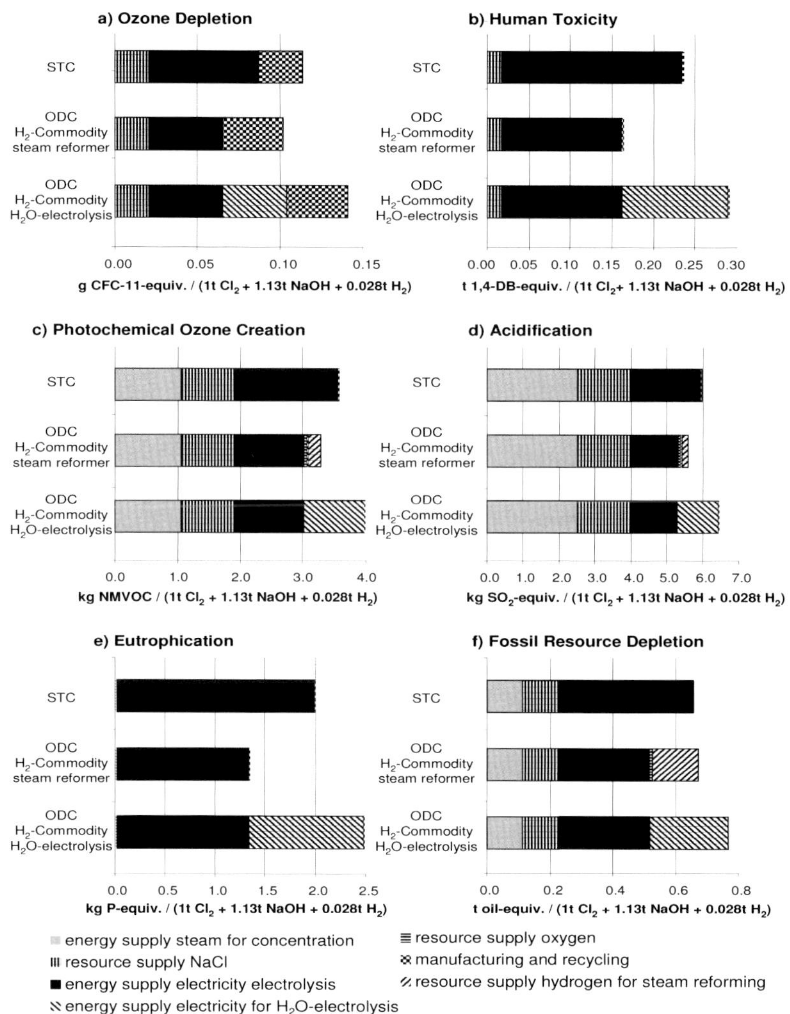

Figure 7.7: Impacts in ozone depletion, human toxicity, photochemical ozone creation, acidification, eutrophication, and fossil resource depletion for scenario H_2-Commodity applying system expansion with added process steam reforming (ODC H_2-Commodity-SteamReformer) and added process H_2O-electrolysis (ODC H_2-Commodity-H_2O-electrolysis), cf. Figure 7.2

The choice between the steam reforming process or the H_2O-electrolysis as added process has a strong influence on the LCIA results (Figures 7.6 and 7.7). Still, the results can also be subject to uncertainties in process data and characterization factors. Therefore, the methods developed in sections 6.2.2.2, 6.2.2.3, and 6.2.2.4 are applied to the LCIA-results of this scenario in section 7.3.2.

7.2.3 Scenario Cl_2-CFP

The total GWI is shown for the STC- and ODC-system in Figure 7.8. These results are equivalent to a CFP of chlorine from a STC- and an ODC-process. The results vary strongly depending on the allocation criteria (Figure 7.8. Chlorine from the ODC-process has a 24 % and 20 % lower CFP than chlorine from the STC-process when mass or market value is used as allocation criterion. In contrast, exergy-based allocation favors chlorine from the STC-process: The CFP for Cl_2 from the STC-process is 28 % higher than the corresponding CFP from the ODC-process. For allocation based on the mole fraction, the CFPs differ by less than 1 %.

The results in Figure 7.8 show only how the choice between the four allocation criteria influences the CFP. Still, it cannot be seen if uncertainty in process data or characterization factors has a similar influence. This influence could be analyzed with a high computational effort using MC-simulations for each result shown in Figure 7.8. Instead, the method derived in section 6.2.3.1 is applied in this work to compare the contribution of uncertainties due to selecting an allocation criteria and due to erroneous process data and characterization factors (section 7.3.3).

7.3 Interpretation and uncertainty analysis

In this section, the LCIA-results of each scenario are interpreted. A sensitivity analysis following section 6.1.4 is applied to identify key parameters. Moreover, uncertainties due to fixing multi-functionality problems are analyzed. All types of uncertainties due to fixing multi-functionality problems (Table 3.2) are present in the three scenarios analyzed in this LCA-study (section 7.1.2). Therefore, all methods proposed in chapter 6 are applied in this section.

In section 7.3.1, the LCIA-results of the scenario H_2-Fuel are discussed. In this scenario, a mix process with a constant mix composition is used as avoided process. Therefore, the uncertainty due to a constant mix composition is analyzed using the method presented in section 6.2.2.1.

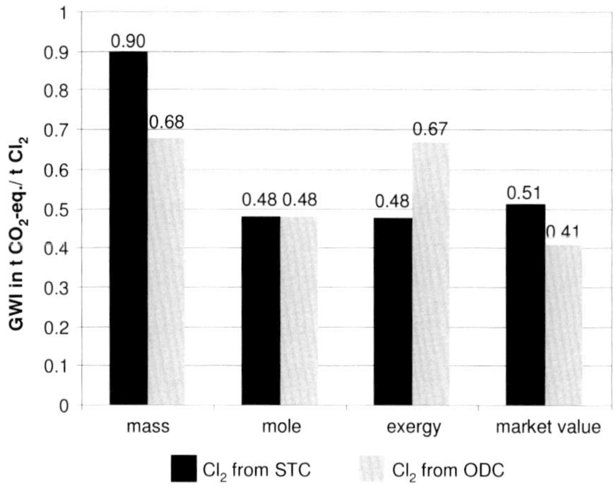

Figure 7.8: GWI for scenario Cl_2-CFP applying allocation based on mass-, mole-, exergy-, and market value-fraction of products

The LCIA-results of the scenario H_2-Commodity are addressed in section 7.3.2. Here, the uncertainty due to the selection between two candidates for the marginal process is analyzed following sections 6.2.2.2 and 6.2.2.3. Moreover, the uncertainty due to additional process data is quantified using the approach shown in section 6.2.2.4.

Finally, the uncertainties in the CFP of Cl_2 (scenario Cl_2-CFP) are the subject of section 7.3.3. The uncertainty due to selecting between the four allocation criteria (Table 7.8) are studied using the method from section 6.2.3.1. Analysis of uncertainty due to constant economic allocation factors following section 6.2.3.2 is also briefly outlined.

The uncertainties in process data (technology matrix $\mathbf{A}_{\mathrm{CA}}$ in Table 7.3) and characterization factors (matrix $\mathbf{QB}_{\mathrm{CA}}$ in Table 7.5) are not available. For illustration, a relative standard deviation σ of 5 % is assumed for each data, i.e.,

$$\sigma((a_{ij})_{\mathrm{CA}}) = 0.05\ a_{ij}, \tag{7.6}$$

and

$$\sigma((qb_{kj})_{\text{CA}}) = 0.05\ (qb)_{kj}. \tag{7.7}$$

7.3.1 Sensitivity and uncertainty analysis of scenario H_2-Fuel

A sensitivity analysis following section 6.1.4 is performed for the LCIA-results of the scenario H_2-Fuel. The analysis focuses on the GWI. The normalized sensitivity coefficients $\frac{\partial h_{\text{GWI}}}{\partial a_{ij}} \frac{a_{ij}}{h_{\text{GWI}}}$ and $\frac{\partial h_{\text{GWI}}}{\partial (qb)_{GW,j}} \frac{(qb)_{GW,j}}{h_{\text{GWI}}}$ for the GWI h_{GWI} with respect to economic flows a_{ij} and characterization factors $(qb)_{GW,j}$ is shown in Table 7.9. Those normalized coefficients refer to the GWI of the STC-process, displayed as the upper bar in Figure 7.4.

The largest sensitivity coefficients for both process data and characterization factors are those with respect to the electricity demand of the STC-process. Moreover, the economic flows corresponding to the products Cl_2, NaOH, and H_2 have also considerable influence on the results. In practice, those parameters are determined by stoichiometry of the electrolysis reaction, cf. Equation 2.1.

The sensitivity coefficients with respect to process data of NaCl supply and steam supply are larger than those with respect to the avoided process combustion NG. But the contribution of both NaCl and steam supply are equal for the GWI of the STC- and ODC-process (Figure 7.4). In contrast, the avoided process affects only the result of the STC-process. Therefore, electricity supply and the avoided process combustion NG are identified as key parameters.

The GWI of both electricity supply and natural gas combustion strongly depend on the local settings: electricity generation varies not only for countries but also for industrial sites. The GWI of natural gas combustions is influenced by the origin and transportation ways of natural gas. Comparing STC- and ODC-processes may have different results depending on the location. The influence of those local settings can be analyzed by varying the GWI of both electricity supply and natural gas combustion.

The total GWI h_{GWI} of the STC- and ODC-system as a function of the GWI of electricity supply and natural gas combustion is shown in Figure 7.9. The GWI of electricity supply $(qb_{1,3})_{\text{CA}}$ is shown on the horizontal axis, the corresponding GWI of natural gas combustion $(qb_{1,14})_{\text{CA}}$ on the vertical axis. The range of $(qb_{1,14})_{\text{CA}}$ is determined by a GWI of combustion NG that was not transported at all and a GWI of combustion NG that includes long gas transportation.

The color axis refers to the total GWI h_{GWI} of the process with a lower impact. The black line marks local settings where STC- and ODC-processes have an equal

GWI. Right of that black line, all combinations of $(qb_{1,3})_{\mathrm{CA}}$ and $(qb_{1,14})_{\mathrm{CA}}$ yield a lower GWI for the ODC-process. In contrast, the GWI of the STC-process is lower for combinations of $(qb_{1,3})_{\mathrm{CA}}$ and $(qb_{1,14})_{\mathrm{CA}}$ left of the black line. Thereby, Figure 7.9 allows graphical identification of beneficial settings for the ODC-process.

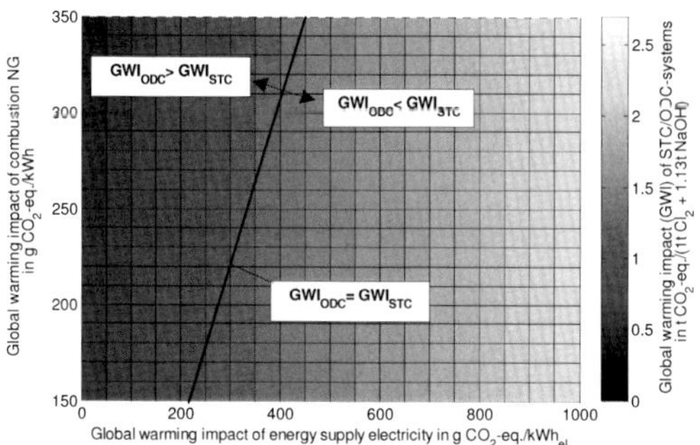

Figure 7.9: Total global warming impact (GWI) h_{GWI} of the STC- and ODC-system as a function of the GWI $(qb_{1,3})_{\mathrm{CA}}$ from electricity supply and the GWI $(qb_{1,14})_{\mathrm{CA}}$ of combustion NG for scenario H_2-Fuel. The colored scale for h_{GWI} refers always to the system with the lower impact, the black line identifies the parameter combination where STC- and ODC-system have an equal GWI.

The avoided process combustion NG is a mix process with a constant mix composition: the corresponding life cycle inventory data uses the German natural gas mix from 2008. The German natural gas mix is composed of natural gas from Germany (GER), the Netherlands (NL), Norway (NO), Russia (RU), and Great Britain (GB) (Bundesverband Energie- und Wasserwirtschaft, 2013). The aggregation factors $v_{j,\mathrm{NG}}$ that correspond with the mix composition of 2008 are shown in Table 7.10. In addition, aggregation factors corresponding with mix compositions between 2010 and 2012 are shown as well. Indeed, the German natural gas mix composition changes over time (Table 7.10).

Following section 6.2.2.1, the uncertainty due to using the constant mix composition from 2008 is analyzed based on average aggregation factors $\tilde{v}_j(\mathrm{NG})$ also shown in Table 7.10. The corresponding variances $\mathrm{var}(\tilde{v}_j(\mathrm{NG}))$ calculated from Equation 6.32 are also displayed in Table 7.10. The variances $\mathrm{var}(\tilde{v}_j(\mathrm{NG}))$ of the aggregation factors differ by more than two orders of magnitude because the share of Russian and

Table 7.9: Normalized sensitivity coefficients $\eta_{GWI} = \frac{\partial h_{GWI}}{\partial a_{ij}} \frac{a_{ij}}{h_{GWI}}$ for the GWI h_{GWI} of the STC-system with respect to process data a_{ij} in the technology $\mathbf{A}_{CA}$ and characterization factors $(qb)_{GWI}$ in the impact matrix $\mathbf{QB}_{CA}$ calculated for the final demand vector $\mathbf{f}_{STC}$

	unit	manufacturing STC plant	manufacturing ODC plant	energy supply electricity	resource supply H_2O	resource supply NaCl	air-fractionation	energy supply steam	STC electrolysis	concentration NaOH, STC	ODC electrolysis	concentration NaOH, ODC	steam reforming	H_2O-electrolysis	combustion NG (mix GER)
STC electrolysis plant	-	0	0	0	0	0	0	0	0	0	0	0	0	0	0
ODC electrolysis plant	-	0	0	0	0	0	0	0	0	0	0	0	0	0	0
electricity	kWh	0	0	-0.78	0	0	0	0	0.78	0	0	0	0	0	0
H_2O	t	0	0	0	0	0	0	0	0	0	0	0	0	0	0
NaCl	t	0	0	0	0	-0.16	0	0	0.16	0	0	0	0	0	0
O_2	t	0	0	0	0	0	0	0	0	0	0	0	0	0	0
N_2	t	0	0	0	0	0	0	0	0	0	0	0	0	0	0
steam	t	0	0	0	0	0	0	-0.17	0	0.17	0	0	0	0	0
$Cl_{2,STC}$	t	0	0	0	0	0	0	0	-0.56	0	0	0	0	0	0
$NaOH_{STC,32}$	t	0	0	0	0	0	0	0	-0.27	0.27	0	0	0	0	0
$NaOH_{STC,50}$	t	0	0	0	0	0	0	0	0	-0.44	0	0	0	0	0
$H_{2,STC}$	t	0	0	0	0	0	0	0	-0.11	0	0	0	0	0	0
$Cl_{2,ODC}$	t	0	0	0	0	0	0	0	0	0	0	0	0	0	0
$NaOH_{ODC,32}$	t	0	0	0	0	0	0	0	0	0	0	0	0	0	0
$NaOH_{ODC,50}$	t	0	0	0	0	0	0	0	0	0	0	0	0	0	0
$H_{2,SR}$	t	0	0	0	0	0	0	0	0	0	0	0	0	0	0
$H_{2,EL}$	t	0	0	0	0	0	0	0	0	0	0	0	0	0	0
$O_{2,EL}$	t	0	0	0	0	0	0	0	0	0	0	0	0	0	0
Natural gas (mix GER)	t H_2-eq.	0	0	0	0	0	0	0	0	0	0	0	0	0	0.11
CO_2-equiv.	t	0	0	0.78	0	0.16	0	0.17	0	0	0	0	0	0	-0.11

Table 7.10: Aggregation factors $v_{j,\text{NG}}$ corresponding with the German natural gas mix composition (Bundesverband Energie- und Wasserwirtschaft, 2013), average aggregation factors $\tilde{v}_{j,\text{NG}}$ and variances of the average aggregation factors $\text{var}(\tilde{v}_{j,\text{NG}})$

	v_{GER}	v_{NL}	v_{NO}	v_{RU}	v_{GB}
v_j(NG 2008)	0.14	0.19	0.26	0.37	0.04
v_j(NG 2010)	0.11	0.22	0.29	0.06	0.33
v_j(NG 2011)	0.13	0.21	0.28	0.07	0.31
v_j(NG 2012)	0.11	0.20	0.23	0.34	0.12
$\tilde{v}_{j,\text{NG}}$	0.123	0.205	0.265	0.21	0.197
$\text{var}(\tilde{v}_{j,\text{NG}})$	1.7E-04	1.3E-04	5.3E-04	2.1E-02	1.5E-02

British natural gas varied considerably more than the shares of German, Dutch, and Norwegian gas.

The average aggregation factors $\tilde{v}_j(\text{NG})$ are used to calculate LCIA-results $h_k(\tilde{v}_j)$ shown in Table 7.11. The overall uncertainty of those LCIA-results is analyzed using Equation 6.26 using input uncertainties for process data (Equation 7.6), characterization factors (Equation 7.7), and aggregation factors (Table 7.10).

The overall uncertainty is expressed in the variance $\text{var}(h_k(\tilde{v}_j))$ and the corresponding relative standard deviation $\frac{\sigma(h_k(\tilde{v}_j))}{h_k(\tilde{v}_j)}$ given in Table 7.11. Additionally, the relative contributions of uncertainty in process data $\zeta(h_k, \mathbf{A}_{\text{CA}})$, in characterization factors $\zeta(h_k, \mathbf{QB}_{\text{CA}})$, and in the constant aggregation factors $\zeta(h_k, \tilde{\mathbf{V}}_{\text{CA}})$ is also shown in Table 7.11.

The relative uncertainty contribution of the aggregation factors is considerably large only for the ozone depletion impact (52.8 %). In all other impact categories, the uncertainty contribution due to a constant natural gas mix composition in Germany is below 5 %. For the impact categories eutrophication and human toxicity, the relative contribution of the aggregation factors is even zero, because the avoided mix process does not contribute environmental impacts to those categories, cf. Figure 7.4. The large uncertainty contribution in ozone depletion may be caused by the fluctuating share of Russian gas in the German natural gas mix. The transportation ways of that gas are longer and subject to potential leakages of gases which contribute to ozone depletion.

Table 7.11: LCIA-results $h_k(\tilde{v}_{j,\mathrm{NG}})$, variances $\mathrm{var}(h_k(\tilde{v}_{j,\mathrm{NG}}))$, relative standard deviation $\frac{\sigma(h_k(\tilde{v}_j))}{h_k(\tilde{v}_j)}$, and relative uncertainty contributions of process data $\zeta(h_k, \mathbf{A}_{\mathrm{CA}})$, characterization factors $\zeta(h_k, \mathbf{QB}_{\mathrm{CA}})$, and aggregation factors $\zeta(h_k, \tilde{\mathbf{V}}_{\mathrm{CA}})$ for the scenario H_2-Fuel

k	$h_k(\tilde{v}_j)$	$\mathrm{var}(h_k(\tilde{v}_j))$	$\frac{\sigma(h_k(\tilde{v}_j))}{h_k(\tilde{v}_j)}$	$\zeta(h_k, \mathbf{A}_{\mathrm{CA}})$	$\zeta(h_k, \mathbf{QB}_{\mathrm{CA}})$	$\zeta(h_k, \tilde{\mathbf{V}}_{\mathrm{CA}})$
CO_2-eq.	2.06 t	$0.028\,t^2$	8.1 %	74.9 %	24.7 %	0.4 %
CFC-11-eq.	0.09 g	$1.3 \times 10^{-4}\,g^2$	12.7 %	35.7 %	11.5 %	52.8 %
SO_2-eq.	5.80 kg	$0.137\,kg^2$	6.4 %	76.6 %	22.5 %	0.9 %
P-eq.	2.00 kg	$0.037\,kg^2$	9.6 %	74.0 %	26.0 %	0.0 %
NMVOC-eq.	3.31 kg	$0.052\,kg^2$	6.9 %	74.4 %	22.3 %	3.3 %
1,4-DB-eq.	0.24 t	$4.7 \times 10^{-4}\,t^2$	9.0 %	74.8 %	25.2 %	0.0 %
oil-eq.	0.56 t	$2.1 \times 10^{-3}\,t^2$	8.2 %	74.8 %	24.9 %	0.3 %

7.3.2 Sensitivity and uncertainty analysis of scenario H_2-Commodity

A sensitivity analysis following section 6.1.4 is also performed for the LCIA-results of the scenario H_2-Commodity. The results are not shown here because they are qualitatively very similar to those displayed in Table 7.9: The GWI is again most sensitive to parameters related to electricity supply.

The total GWI h_{GWI} is shown as a function of the GWI of electricity supply in Figure 7.10. The GWI of electricity supply $(qb_{13})_{\mathrm{CA}}$ (Table 7.5) is shown on the horizontal axis. The black line refers to the GWI of the STC-process. The dashed line represents the GWI from the ODC-process system expanded with a steam reforming process, and the dotted line the GWI from the ODC-process system expanded with H_2O-electrolysis. The dotted black line marks the GWI of electricity supply in Germany (Table 7.2 and 7.5) which corresponds with the results shown in Figure 7.6.

The GWI of the ODC-process with a steam reforming process intersects with the GWI of the STC-process at an GWI of electricity supply of 443 g CO_2/kWh. The STC-process has thus an overall lower GWI at locations where electricity supply has a GWI lower than 443 g CO_2/kWh (cf. Figure 7.10). In contrast, the combination of ODC-process and a steam reforming process has a lower overall GWI at locations where electricity supply has a GWI higher than 443 g CO_2/kWh. That is likely in countries where electricity generation is based on fossil fuels.

Chlor-alkali electrolysis plants are mainly operated in countries with large chemical industries. Apart from Germany, other countries with large chemical industries also

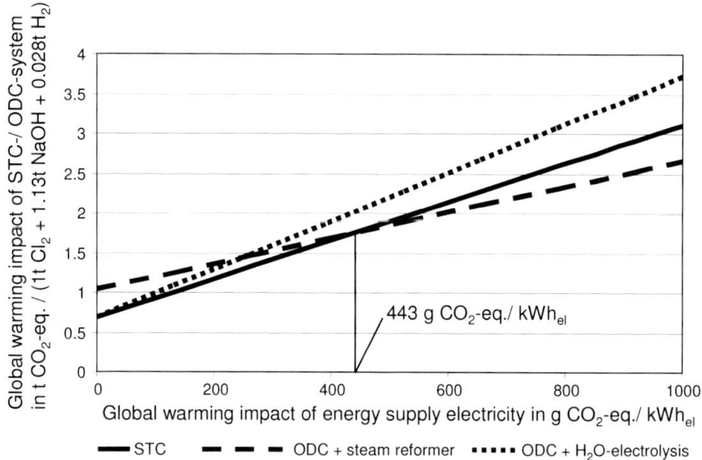

Figure 7.10: Total global warming impact (GWI) h_{GWI} of the STC- and ODC-system as a function of the GWI $(qb_{\mathrm{GWI}})_{\mathrm{CA}}$ of electricity supply for scenario H_2-Commodity

rely on electricity generation from fossil fuels, e.g., China, USA, or Japan. Electricity generation in those countries has a similar GWI as that in Germany. The ODC-process has thus a realistic potential to reduce worldwide GWI in the chemical industry even if hydrogen is used as chemical commodity.

The ODC-process system expanded with a H_2O-electrolysis has always a higher GWI than the STC-process. The combination of ODC- and H_2O-electrolysis is thus not favorable from a global warming perspective.

The uncertainty due to selecting between a steam reforming process and H_2O-electrolysis as added process can be analyzed using the approach presented in section 6.2.2.2. Here, the simplified equations for aggregation of $\alpha = 2$ unit processes are applicable (section 6.2.2.3).

The average aggregation factors for $\alpha = 2$ unit processes (steam reforming and H_2O-electrolysis) become $\tilde{v}_{\text{steam reforming}} = \tilde{v}_{\mathrm{H_2O\text{-}electrolysis}} = \frac{1}{\alpha} = 0.5$ (cf. Equation 6.34). The GWI of the ODC-system corresponding to the average aggregation factors becomes $h_{\mathrm{GWI}} = 2.39\,\mathrm{t}\ \mathrm{CO}_{2_{\mathrm{eq}}}$.

The overall uncertainty of that GWI is $\mathrm{var}(h_{\mathrm{GWI}}) = 0.128\,\mathrm{t}^2$ calculated from Equation 6.26 using input uncertainties for process data (Equation 7.6), characterization factors (Equation 7.7), and average aggregation factors (Equations 6.41 and 6.43). This overall uncertainty corresponds to a relative standard deviation $\frac{\sigma(h_{\mathrm{GWI}})}{h_{\mathrm{GWI}}} = 15\,\%$. Applying Equation 6.44, the relative contribution $\zeta(h_{\mathrm{GWI}}, \mathbf{V}_{\mathrm{CA}})$ of the uncertainty due to selecting between the steam reforming process and the H_2O-electrolysis to the overall uncertainty is $\zeta(h_{\mathrm{GWI}}, \mathbf{V}_{\mathrm{CA}}) = 70\,\%$. The selection of the added process has thus not only considerable influence on the LCIA-results (Figures 7.6 and 7.7), the contribution of the uncertainty due to this selection also outweighs uncertainties in process data and characterization factors.

The combination of ODC- and H_2O-electrolysis is used for illustration of the analysis of uncertainties due to additional process data from system expansion (section 6.2.2.4). System expansion for H_2O-electrolysis uses additional process data in the column H_2O-electrolysis of the technology matrix $\mathbf{A}_{\mathrm{CA}}$ on p. 118. The additional process data is denoted $\mathbf{A}"_{\mathrm{CA}}$.

The uncertainty of the LCIA-results h_k are expressed in the total variances $\mathrm{var}(h_k)$ and the corresponding relative standard deviation $\frac{\sigma(h_k)}{h_k}$ calculated from Equation 6.26 and shown in Table 7.12. The relative uncertainty contribution from regular process data and characterization factors $\zeta(h_k, \mathbf{A}_{\mathrm{CA}}, \mathbf{QB}_{\mathrm{CA}})$ and the relative contribution from the additional process data $\zeta(h_k, \mathbf{A}"_{\mathrm{CA}})$ is calculated following section 6.2.2.4 and also given in Table 7.12.

Table 7.12: LCIA-results h_k, total variance $\mathrm{var}(h_k)$, relative standard deviation $\frac{\sigma(h_k)}{h_k}$, and relative uncertainty contributions $\zeta(h_k, \mathbf{A}_{\mathrm{CA}}, \mathbf{QB}_{\mathrm{CA}})$ of regular process data and characterization factors, and relative uncertainty contributions $\zeta(h_k, \mathbf{A}"_{\mathrm{CA}})$ of added process data $\zeta(h_k, \mathbf{A}"_{\mathrm{CA}})$ calculated for system expansion with the H_2O-electrolysis in the scenario H_2-Commodity

k	h_k	$\mathrm{var}(h_k)$	$\frac{\sigma(h_k)}{h_k}$	$\zeta(h_k, \mathbf{A}_{\mathrm{CA}}, \mathbf{QB}_{\mathrm{CA}})$	$\zeta(h_k, \mathbf{A}"_{\mathrm{CA}})$
CO_2-eq.	2.66 t	$0.033\,\mathrm{t}^2$	6.8 %	87.3 %	12.7 %
CFC-11-eq.	0.14 g	$0.0001\,\mathrm{g}^2$	7.1 %	91.0 %	9.0 %
SO_2-eq.	6.43 kg	$0.121\,\mathrm{kg}^2$	5.4 %	94.9 %	5.1 %
P-eq.	2.47 kg	$0.044\,\mathrm{kg}^2$	8.5 %	85.5 %	14.5 %
NMVOC-eq.	3.97 kg	$0.054\,\mathrm{kg}^2$	5.9 %	92.0 %	8.0 %
1,4-DB-eq.	0.29 t	$0.001\,\mathrm{t}^2$	10.9 %	86.0 %	14.0 %
oil-eq.	0.76 t	$0.003\,\mathrm{t}^2$	7.2 %	88.1 %	11.9 %

In this example, the relative uncertainty contribution due to added process data is between 5.1 % (impact category acidification) and 14.5 % (impact category eutrophication). However, these results are based on the assumed process data uncertainties and therefore purely for illustration of the method.

7.3.3 Uncertainty analysis of scenario Cl_2-CFP

The CFP of chlorine from the STC-process is already known to depend on the chosen allocation criterion (Leimkühler et al., 2010). The results shown in Figure 7.8 confirm this dependency for the CFP of chlorine from the ODC-process. The method derived in section 6.2.3.1 can be used to systematically quantify the uncertainty contribution of the choice between the four allocation criteria assessed in the scenario Cl_2-CFP (Table 7.8).

The average allocation factors based on the four criteria assessed in section 7.1.5.3 for both STC- and ODC-processes (Table 7.8) are shown in Table 7.13. The variances $\mathrm{var}(\tilde{c})$ of the average allocation factors are calculated from Equation 6.47 and also given in Table 7.8.

Table 7.13: Average allocation factor $\tilde{c}$ and variance of the average allocation factors $\mathrm{var}(\tilde{c})$ for both the STC- and ODC-process

criteria	STC			ODC	
	Cl_2	NaOH	H_2	Cl_2	NaOH
$\tilde{c}$	0.307	0.494	0.199	0.389	0.611
$\mathrm{var}(\tilde{c})$	0.0084	0.0188	0.0318	0.0067	0.0067

The carbon footprints h_{GWI} of Cl_2 from the STC- and ODC-systems are calculated for the average allocation factor. The overall uncertainty expressed in the total variance $\mathrm{var}(h_{\mathrm{GWI}})$ and the relative standard deviation $\frac{\sigma(h_{\mathrm{GWI}})}{h_{\mathrm{GWI}}}$ are calculated from Equation 6.26. For this overall uncertainty, the relative contributions of uncertainty in process data and characterization factors $\zeta(h_k, \mathbf{A}_{\mathrm{CA}}, \mathbf{QB}_{\mathrm{CA}})$ and in the average allocation factor $\zeta(h_{\mathrm{GWI}}, \tilde{\mathbf{C}}_{\mathrm{CA}})$ are shown in Table 7.14.

The carbon footprints h_{GWI} of Cl_2 are almost equal for STC- and ODC-processes but the total variances $\mathrm{var}(h_{\mathrm{GWI}})$ are not. The CFP of Cl_2 from STC-process is subject to more uncertainty from allocation factors. The reason is the third and non-common product hydrogen, which causes a greater variation between the allocation factors. The contribution from uncertainty due to the choice between four allocation criteria dominates uncertainties due to process data and characterization factors because $\zeta(h_{\mathrm{GWI}}, \mathbf{C}_{\mathrm{CA}}) > 95\,\%$ for both CFPs.

For economic allocation, constant allocation factors based on the market prices of Cl_2, NaOH, and H_2 are used (Table 7.8). The uncertainty due to these constant economic allocation factors can be analyzed following the approach in section 6.2.3.2. However, this uncertainty is not further elaborated in this work because historic market price for H_2 were not available.

Table 7.14: LCIA-results h_k, total variance $\mathrm{var}(h_k)$, relative standard deviation $\frac{\sigma(h_{\mathrm{GWI}})}{h_{\mathrm{GWI}}}$, relative uncertainty contributions of process data and characterization factors $\zeta(h_k, \mathbf{A}_{\mathrm{CA}}, \mathbf{QB}_{\mathrm{CA}})$, and allocation factors $\zeta(h_k, \tilde{\mathbf{C}}_{\mathrm{CA}})$ calculated for the average allocation factor $\tilde{c}$ in scenario Cl_2-CFP

	h_{GWI} in t CO_2-equiv.	$\mathrm{var}(h_{\mathrm{GWI}})$ in $(t\ CO_2\text{-equiv.})^2$	$\frac{\sigma(h_{\mathrm{GWI}})}{h_{\mathrm{GWI}}}$	$\zeta(h_{\mathrm{GWI}}, \mathbf{A}_{\mathrm{CA}}, \mathbf{QB}_{\mathrm{CA}})$	$\zeta(h_{\mathrm{GWI}}, \mathbf{C}_{\mathrm{CA}})$
STC	0.59 t	0.127 t^2	60.4 %	2.2 %	97.8 %
ODC	0.56 t	0.056 t^2	42.3 %	3.7 %	96.3 %

7.4 Discussion and outlook

The results of the LCA-study comparing the STC- and ODC-process favor the ODC-technology in most scenarios and impact categories (Figures 7.4 to 7.8). The results are subject to uncertainties due to fixing multi-functionality problems. Indeed, the contribution of these uncertainties are large compared to uncertainties from process data and characterization factors (section 7.3.2 and Table 7.14). But this interpretation has to be interpreted with caution because the uncertainty in process data and characterization factors are assumptions. Still, it is unlikely that the actual uncertainties in process data and characterization factors are high enough to outweigh the uncertainty in selecting an allocation criterion in scenario Cl_2-CFP.

The scenarios H_2-Fuel and H_2-Commodity are likely to be used for decision making in the chlor-alkali industry. In both scenarios, the ODC-process has an environmental advantage in most analyzed impact categories. Still, a decision for one or the other technology or even for a retrofit of the ODCs must consider the economic settings. A production site that utilizes H_2 as fuel (scenario H_2-Fuel) seems to be better qualified because using natural gas instead of H_2 will most likely not require additional investment costs. But if H_2 is used as a chemical commodity, steam reforming as added process is not necessarily existing at the production site and thus requires additional investment costs. In a future study, economic settings of the chlor-alkali industry could be integrated by including market effects into an LCA-study.

The comparative LCA-study presented in this chapter included all types of uncertainty due to fixing multi-functionality problems (Table 3.2). These uncertainties are successfully analyzed and compared to uncertainties in process data and characterization factors using the methods presented in chapter 6. The uncertainty analyses in sections 7.3.1 to 7.3.3 were performed quickly using the matrix formulation from chapter 4 and Equation 6.26: The calculations took a few seconds ($< 10\,\mathrm{s}$) in Matlab using a dual core processor with 2.33 GHz. In future work, these figures should be compared to MC-simulations as it was done by Imbeault-Tétreault et al. (2013) for uncertainties in process data and characterization factors.

The chlor-alkali process system used in this work contained 14 unit processes and 19 economic flows. This dimension is rather small compared to other LCA-studies that focus on uncertainty analysis: Imbeault-Tétreault et al. (2013) studies a process system with 881 unit processes; Ciroth et al. (2004) uses a virtual process system with 60 unit processes. Therefore, the approach presented in this work should be tested with a larger process system in the future.

The results of the uncertainty analysis are calculated in Matlab. An application of the methods to larger process systems will likely become inconvenient using only Matlab. Therefore, it is desirable that the formulas derived in this work are implemented in existing software tools. Suitable tools for an implementation are CMLCA (Heijungs, 2013a) or Simapro (PRé Consultants, 2013) because they are both based on the matrix formulation. Those two tools already include algorithms for MC-simulations. They are thus predestined to execute a comparison between the approach presented in this work and MC-simulations.

Chapter 8

Conclusions

8.1 Conclusions for the chlor-alkali industry

Chlor-alkali electrolysis is known as an energy-intensive industrial process. Novel ODCs reduce the electricity demand of chlor-alkali electrolysis by 30 %. In this work, the potential environmental impacts of chlor-alkali electrolysis with ODCs are investigated and compared with those of existing electrolysis plants. The comparative LCA-study reveals that ODC electrolysis indeed reduces environmental impacts compared to membrane electrolysis without ODCs. However, the reductions are typically smaller than the 30 % reductions in electricity demand. The discrepancy in reductions of electricity demand and environmental impacts results mainly from the non-common co-product hydrogen. Hydrogen is only co-produced by membrane electrolysis but not by ODC electrolysis. In contrast to comparing electricity demand, the holistic LCA-based comparison includes utilization of hydrogen from membrane electrolysis.

In this comparative LCA-study, three scenarios for the functional unit are analyzed. These three scenarios represent site-specific settings for chlor-alkali electrolysis. The two scenarios H_2-Fuel and H_2-Commodity represent two alternative utilizations of hydrogen. They differ in fixing the multi-functionality problem during the comparative LCA-study: If H_2 is used as a fuel, avoided environmental impacts of natural gas combustion are subtracted from the impacts of the membrane electrolysis. If H_2 is used as a commodity, environmental impacts from a steam reforming process are added to the impacts from the ODC electrolysis. Both scenarios yield by 12 % (H_2-Fuel) and 8 % (H_2-Commodity) less global warming impact for the ODC electrolysis.

The third scenario studied refers to the carbon footprint of chlorine from a membrane and an ODC electrolysis. In this scenario, the global warming impact has to be allocated between the co-products of both membrane and ODC electrolysis. The carbon footprint, however, depends strongly on the allocation criterion: It is lower for chlorine from the ODC electrolysis if allocation is based on the mass-fraction and on the market value. But chlorine from the ODC electrolysis has a higher carbon footprint if allocation is based on exergy. In general, both technologies should not be compared based on the carbon footprint of chlorine because the result depends strongly on the selected allocation criterion. Instead, the comparison of technologies should rather be based on the scenarios H_2-Fuel and H_2-Commodity.

The sensitivity analysis revealed that environmental impacts from electricity supply influences the results considerably. The presented figures are valid for chlor-alkali electrolysis plants operated in Germany. In fact, electricity generation varies from country to country. Therefore it is very important to adapt the environmental impacts of electricity supply if the results are transferred to other countries. For global warming impact, ODC electrolysis is favorable in countries where electricity generation is based on fossil fuels. Most chlor-alkali electrolysis plants are located in countries with large chemical industries, e.g., China, USA, Germany, or Japan. The electricity generation of those countries is mainly based on fossil fuels. The novel ODC technology thus has potential to reduce worldwide global warming impact.

8.2 Conclusions for LCA methodology

Multi-product processes can cause multi-functionality problems in LCA. Methods for fixing multi-functionality problems are available. So far, the systematic application of those methods remained an open question for comparative LCA of multi-product processes. In this work, a procedure is presented that recommends how multi-functionality problems should be fixed in comparative LCA of multi-product processes. The procedure is specified in a workflow that is conform with the existing ISO-standard.

The workflow is based on the distinction of main products and by-products from the multi-product process. Based on that distinction, the choice between system expansion, avoided burden, and allocation is eliminated: Multi-functionality problems due to non-common main products should be fixed with system expansion. Multi-functionality problems due to non-common by-products should preferably be fixed with avoided burden, but allocation can be necessary if no avoided process exists or endless loops of by-products occur. The workflow is applicable if the economic settings are known so that main products and by-products can be distinguished. Remaining choices exist only within each method but not between them.

The ISO-standard requires LCA practitioners to conduct and document a sensitivity analysis for the choices common to system expansion, avoided burden, and allocation. This work presents a systematic method for analyzing sensitivities and uncertainties due to fixing multi-functionality problems with system expansion, avoided burden, and allocation. The method is based on earlier work that applied analytical error propagation to study uncertainties in process data and characterization factors in LCA. In this work, that approach is expanded for uncertainties due to fixing multi-functionality problems. For this purpose, the existing matrix formulas for calculating LCA-results are expanded by explicit formulas modeling system expansion, avoided burden, and allocation. The uncertainties due to fixing multi-functionality problems are modeled as uncertain parameters included in those formulas.

The presented method allows for a systematical and automated uncertainty analysis using a single equation. This holistic approach is an advantage compared to existing uncertainty analysis tools. It identifies processes and parameters strongly influenced by fixing multi-functionality problems and those dominated by uncertainties in process data or characterization factors. The latter are predestined for further data refinement.

The formulas derived in this work could be implemented in LCA-software tools. Thereby the method would be available to a large number of LCA-practitioners. Moreover, it could be tested whether the approach requires indeed less computational effort than the frequently used Monte-Carlo simulation. Nevertheless, Monte-Carlo simulations are capable to calculate nearly exact uncertainties while the analytical approach is based on a first-order approximation. However, it is shown that the first-order approximation is exact for propagation of uncertainties in allocation factors, if no loops of the allocated functional flows exist. Still, future work should study the computational advantage and the potential loss of information of the analytical approach compared to Monte-Carlo simulations in cases where the first-order approximation is not exact.

Both the procedure for fixing multi-functionality problems in comparative LCA and the analytical approach for analyzing uncertainties due to fixing multi-functionality problems complement the ISO-standard. Thereby, this work improves the applicability of LCA to compare industrial processes. Decision makers who apply the systematic approach can now study all types of uncertainty in comparative LCA of industrial processes. Decisions for particular industrial processes to reduce environmental impacts will ultimately become more reliable.

Glossary

added process

an extra unit process added to a process system during system expansion in comparative LCA 28

aggregation factor

factor describing how multiple processes are aggregated into a single process 50

aggregation matrix

matrix for aggregating multiple processes into a single process used during avoided burden or system expansion 51

allocation

fixing multi-functionality problems by dividing multi-functional unit process into mono-functional unit processes 31

allocation criterion

a relationship between input and output flows of a unit process that allows deriving an allocation factor 32

allocation factor

factor for allocating non-functional economic flows and elementary flows of a multi-functional unit process to a mono-functional unit processes 31

allocation matrix

matrix for allocating non-functional flows of a multi-functional unit process to mono-functional unit processes during allocation 58

alternative processes

multiple processes producing alternative products satisfying the same functional unit 17

alternative products

multiple products satisfying the same functional unit 17

avoided burden

fixing multi-functionality problems by subtracting avoided environmental impacts from a multi-functional process system 29

avoided process

a unit process used to calculate avoided environmental impacts during fixing multi-functionality problems with avoided burden 29

by-product

product that has a positive economic value but is not economic reason of process operation 71

characterization factor

factor that converts an elementary flow into an environmental impact (ISO (2006a) on p. 12) 21

characterization matrix

a matrix comprising the characterization factors for all elementary flows of a process system 22

combined waste processing

a unit process with two or more waste input flows 23

common functional flow

functional flow from alternative processes that are equivalent in properties and quantities of the represented energy or material flow 70

comparative LCA

a LCA study aiming to compare environmental impacts of alternative products or alternative processes 17

discrepancy matrix

a matrix comprising discrepancy vectors of alternative products or processes 67

discrepancy vector

the difference between final supply vector and final demand vector 26

economic allocation

allocation criterion based on economic figures such as market prices 33

economic flow

a material or energy flow exchanged between unit processes or process systems 17

elementary flow

a material or energy flow exchanged between unit processes and the environment (ISO (2006a) on p. 9) 16

equivalence matrix

matrix for adding rows of equivalent economic flows used during avoided burden or system expansion 48

equivalent economic flow

economic flows that can perform the same function 30

final demand matrix

a matrix comprising final demand vectors of alternative products or processes 66

final demand vector

a vector representing the functional unit by comprising all functional flows 20

final supply matrix

a matrix comprising final demand vectors of alternative products or processes 67

final supply vector

a vector representing the functional unit by comprising all functional flows 26

function matrix

matrix for creating mono-functional unit processes during allocation by deleting functional flows in the columns of a multi-functional unit process and deleting non-functional flows in the columns of duplicated unit processes 56

functional flow

economic flows that corresponds with a process' function. Functional flows of industrial processes are either product output or waste input flows 17

functional unit

quantified performance of a process system for use as a reference unit, expressed in functional flows (ISO (2006a) on p. 10) 17

goal and scope definition

first step in LCA in which the intended application and underlying conditions are stated 15

impact assessment

third step in LCA in which inputs and outputs of a process system are associated with environmental impacts (ISO (2006a) on p. 7) 15

interpretation

fourth step in LCA in which the results of the inventory analysis and impact assessment are discussed in the context of goal and scope definition (ISO (2006a) on p. 7) 15

intervention matrix

a matrix comprising all elementary flows of all unit process within a process system 20

inventory analysis

second step in LCA in which data is collected and the inputs and outputs of a process system are calculated (ISO, 2006a) 15

LCA-result

scaling vector **s**, total intervention vector **g**, and total impact vector **h** as results of life cycle assessment 22

LCI-result

scaling vector **s** and total intervention vector **g** as result of the life cycle inventory analysis (ISO (2006a) on p. 10) 21

LCIA-result

total impact vector **h** as result of life cycle impact assessment 22

life cycle assessment

evaluation of potential environmental impacts throughout a product's life cycle (ISO (2006a) on p. 7) 14

main product

product that is economic reason of process operation 71

marginal process

unit process that provides an extra demand of a product used during system expansion and avoided burden 29

mix process

an aggregated process representing the market mix of unit processes for production of a product 29

mono-functional

a unit process with only one functional flow 23

multi-functional

a unit process with two or more functional flows 23

multi-functional process system

a process system that contains at least one multi-functional unit process 24

multi-functionality problem

a multi-functionality problem occurs whenever a process system generates a set of equations that cannot be uniquely solved for calculation of LCA-results 24

multi-product process

a unit process with two or more product output flows 23

physical relationship

a relationship that reflects the way inputs and outputs of a unit process are changed by a quantitative change in the functional flows of the unit process 32

process matrix

a matrix comprising all process vectors of a process system 20

process system

set of unit processes describing parts or an entire product life cycle 16

process vector

a vector comprising all input and output flows of a unit process 19

product

an economic flow with a positive economic value (ISO (2006a) on p. 8) 17

pseudoinverse

a generalized inverse for non-square and singular matrices, e.g. defined by Moore (1920) and Penrose (1955) 25

recycling process

a unit process with at least one product output flow and at least one waste input flow 23

reference flow

measure of outputs from processes in a given system required to fulfill the functional unit, in this work equivalent to the functional unit (ISO (2006a) on p. 11) 17

scaling matrix

a matrix comprising scaling vectors of alternative products or processes 66

scaling vector

a vector comprising scaling factor for every unit process of a process system 21

single-entry-matrix

matrix whose elements are all zero except for a one non-zero element 85

system expansion

fixing multi-functionality problems by expanding the functional unit for additional functional flows 27

technology matrix

a matrix comprising all economic flows of all unit process within a process system 20

total impact matrix

a matrix comprising total impact vectors of alternative products or processes 66

total impact vector

a vector comprising all environmental impacts associated with a process system and a given final demand vector 22

total intervention matrix

a matrix comprising total intervention vectors of alternative products or processes 66

total intervention vector

a vector comprising all elementary flows associated with a process system and a given final demand vector 21

transformation matrix

matrix for duplicating columns of multi-functional unit processes during allocation 55

unit process

model of an material or energy transformation quantified with input and output data (ISO (2006a) on p. 12) 16

waste

an economic flow with a negative economic value (ISO (2006a) on p. 12) 17

Appendix A

Proof of linear relationship between LCA-results and allocation factors

A mathematical proof is presented for the following statement:
LCA-results depend linear on allocation factors if the allocated functional flows are not input flows of unit processes that belong to the life cycle of the allocated functional flows (i.e., no loops of allocated functional flows).

A fixed technology matrix with no loops of allocated functional flows is denoted by $\mathbf{A}_{\text{fixed,nl}}$ and can be written as

$$\mathbf{A}_{\text{fixed, no loop}} = \left(\begin{array}{c|c} \mathbf{A}_{\text{upstream}} & \overbrace{\mathbf{A}_{\text{allocated}}(c_{pq})}^{=a_{ij}\cdot c_{pq}} \\ \hline \mathbf{A}_{\text{no loop}} & \mathbf{A}_{\text{functional}} \end{array} \right). \tag{A.1}$$

Here, the submatrices $\mathbf{A}_{\text{allocated}}$ and $\mathbf{A}_{\text{functional}}$ contain the allocated mono-functional unit processes. The submatrix $\mathbf{A}_{\text{allocated}}$ includes the allocated non-functional flows and the submatrix $\mathbf{A}_{\text{functional}}$ the functional flows. The allocated non-functional flows in the matrix $\mathbf{A}_{\text{allocated}}$ depend linearly on the allocation factors c_{pq} (cf. Equations 3.28 and 6.6). The functional flows in the submatrix $\mathbf{A}_{\text{functional}}$ form a square diagonal matrix.

The submatrices $\mathbf{A}_{\text{upstream}}$ and $\mathbf{A}_{\text{no loop}}$ contain unit processes not involved in the allocation procedure. The submatrix $\mathbf{A}_{\text{upstream}}$ includes the upstream economic flows required by the allocated mono-functional unit processes. The submatrix $\mathbf{A}_{\text{no loop}}$ contains the functional flows of the allocated mono-functional unit processes. This submatrix $\mathbf{A}_{\text{no loop}}$ contains only zeroes due to the assumption that no loops of the allocated functional flows exist. Therefore, it is

$$\mathbf{A}_{\text{fixed, no loop}} = \left(\begin{array}{c|c} \mathbf{A}_{\text{upstream}} & \overbrace{\mathbf{A}_{\text{allocated}}(c_{pq})}^{=a_{ij}\cdot c_{pq}} \\ \hline \mathbf{0} & \mathbf{A}_{\text{functional}} \end{array} \right). \tag{A.2}$$

The fundamental inventory calculation (Equation 3.6) can thus be written as

$$\left(\begin{array}{c|c} \mathbf{A}_{\text{upstream}} & \overbrace{\mathbf{A}_{\text{allocated}}(c_{pq})}^{=a_{ij}\cdot c_{pq}} \\ \hline \mathbf{0} & \mathbf{A}_{\text{functional}} \end{array} \right) \cdot \left(\begin{array}{c} \mathbf{s}_{\text{allocated}} \\ \hline \mathbf{s}_{\text{functional}} \end{array} \right) = \left(\begin{array}{c} \mathbf{f}_{\text{allocated}} \\ \hline \mathbf{f}_{\text{functional}} \end{array} \right). \tag{A.3}$$

This matrix product can be rearranged (cf. Magnus and Neudecker (2007) on p. 12) to

$$\left(\begin{array}{c} \mathbf{A}_{\text{upstream}}\mathbf{s}_{\text{allocated}} + \mathbf{A}_{\text{allocated}}(c_{pq})\mathbf{s}_{\text{functional}} \\ \hline \mathbf{0}\mathbf{s}_{\text{allocated}} + \mathbf{A}_{\text{functional}}\mathbf{s}_{\text{functional}} \end{array} \right) = \left(\begin{array}{c} \mathbf{f}_{\text{allocated}} \\ \hline \mathbf{f}_{\text{functional}} \end{array} \right). \tag{A.4}$$

The upper and lower equation systems in Equation A.4 can be solve independently: The final demand sub-vectors $\mathbf{f}_{\text{allocated}}$ and $\mathbf{f}_{\text{functional}}$ are both given; the submatrix $\mathbf{A}_{\text{functional}}$ is square and diagonal. Thus, the scaling sub-vector $\mathbf{s}_{\text{functional}}$ is not a function of the allocation factors c_{pq} and obtained from

$$\mathbf{s}_{\text{functional}} = \left(\mathbf{A}_{\text{functional}}\right)^{-1} \mathbf{f}_{\text{functional}}. \tag{A.5}$$

Inserting $\mathbf{s}_{\text{functional}}$ in the upper equation system allows calculating the scaling subvector $\mathbf{s}_{\text{allocated}}$ from

$$\mathbf{s}_{\text{allocated}} = \overbrace{\left(\mathbf{A}_{\text{upstream}}\right)^{-1}}^{\neq f(c_{pq})} \cdot \left(\overbrace{\mathbf{f}_{\text{allocated}}}^{\neq f(c_{pq})} - \overbrace{\mathbf{A}_{\text{allocated}}(c_{pq})}^{=a_{ij}\cdot c_{pq}} \cdot \overbrace{\left(\mathbf{A}_{\text{functional}}\right)^{-1} \mathbf{f}_{\text{functional}}}^{\neq f(c_{pq})} \right). \tag{A.6}$$

Here, the scaling sub-vector $\mathbf{s}_{\text{allocated}}$ depends only linearly on the allocation factor c_{pq} because only the submatrix $\mathbf{A}_{\text{allocated}}$ depends linearly on the allocation factor.

Overall, the scaling vector $\mathbf{s} = \left(\begin{array}{c} \mathbf{s}_{\text{allocated}}(c_{pq}) \\ \hline \mathbf{s}_{\text{functional}} \end{array} \right)$ depends only linearly on the allocation factor c_{pq} if no loops of the allocated functional flows exist. The sensitivity coefficient of a scaling vector elements s_j with respect to allocation factors c_{pq} (cf. Equation 6.13) is thus constant:

$$\frac{\partial s_j}{\partial c_{pq}} = \text{const.} \tag{A.7}$$

The corresponding fixed intervention matrix $\mathbf{B}_{\text{fixed}}$ and can be written as

$$\mathbf{B}_{\text{fixed}} = \left(\mathbf{B}_{\text{upstream}} \;\middle|\; \mathbf{B}_{\text{allocated}}(c_{pq}) \right). \tag{A.8}$$

The submatrix $\mathbf{B}_{\text{allocated}}$ depends also linearly on the allocation factor c_{pq}. Following Equation 4.41, the total intervention vector $\mathbf{g}$ is thus calculated from

$$\begin{aligned} \mathbf{g} &= \left(\mathbf{B}_{\text{upstream}} \;\middle|\; \mathbf{B}_{\text{allocated}}(c_{pq}) \right) \cdot \left(\frac{\mathbf{s}_{\text{allocated}}(c_{pq})}{\mathbf{s}_{\text{functional}}} \right) \\ &= \left(\mathbf{B}_{\text{upstream}} \cdot \mathbf{s}_{\text{allocated}}(c_{pq})\right) + \left(\mathbf{B}_{\text{allocated}}(c_{pq}) \cdot \mathbf{s}_{\text{functional}}\right). \end{aligned} \tag{A.9}$$

Here, the total intervention vector $\mathbf{g}$ depends in two terms linearly on the allocation factor c_{pq}. Therefore, the corresponding sensitivity coefficient of a total intervention vector element g_i is

$$\frac{\partial g_i}{\partial c_{pq}} = \text{const.} \tag{A.10}$$

Following Equation 4.42, the total impact vector $\mathbf{h}$ depends also linearly on the c_{pq}. The corresponding sensitivity coefficients are thus also constant.

For linear functions, the propagation of uncertainties with a first-order approximation (Equation 6.24) is exact (cf. Morgan and Henrion (1992) on p. 186). Therefore, the linear dependence between LCIA-results and allocation factors implies that uncertainties in allocation factors are exactly propagated into uncertainties in LCIA-results $\mathbf{h}$ (cf. Equation 6.26) if no loops of allocated functional flows exist.

The example discussed in section 4.2 illustrates the previous considerations. The product of the matrices $(\mathbf{U} \circ (\mathbf{AT}_1))_{\text{CHP}}$ (Equation 4.30) and $\mathbf{C}_{\text{CHP}}$ (Equation 4.32) yields the fixed technology matrix $\mathbf{A}_{\text{CHP, no loop}}$ as a function of the allocation factors $c_{\text{electricity}}$ and c_{heat}, i.e.,

$$\mathbf{A}_{\text{CHP, no loop}} = (\mathbf{U} \circ (\mathbf{AT}_1))_{\text{CHP}} \cdot \mathbf{C}_{\text{CHP}}$$

$$
= \left(\begin{array}{c|cccc} 1 & -0.42 \cdot c_{\text{electricity}} & -0.42 \cdot c_{\text{heat}} & -0.15 & -0.13 \\ \hline 0 & 1 & 0 & 0 & 0 \\ 0 & 0 & 1.67 & 0 & 0 \\ 0 & 0 & 0 & 1 & 0 \\ 0 & 0 & 0 & 0 & 1 \end{array}\right) \tag{A.11}
$$

The corresponding intervention matrix (cf. Equation 4.3) also depends on these allocation factors, i.e.,

$$
\begin{aligned}
\mathbf{B}_{\text{CHP, no loop}} &= (\mathbf{BT}_0)_{\text{CHP}} \cdot \mathbf{C}_{\text{CHP}} \\
&= \left(\begin{array}{c|cccc} -1.2 & 0 & 0 & 0 & 0 \\ 0.73 & 1.54 \cdot c_{\text{electricity}} & 1.54 \cdot c_{\text{heat}} & 0.55 & 0.48 \end{array}\right)
\end{aligned} \tag{A.12}
$$

Following Equations 4.40 to 4.42, the scaling vector $\mathbf{s}_{\text{CHP, no loop}}$, the total intervention vector $\mathbf{g}_{\text{CHP, no loop}}$, and the total impact vector $\mathbf{h}_{\text{CHP, no loop}}$ depend linearly on the allocation factor $c_{\text{electricity}}$, i.e.,

$$
\begin{aligned}
\mathbf{s}_{\text{CHP, no loop}} &= \mathbf{A}^{-1}_{\text{CHP, no loop}} \cdot \mathbf{f}_{\text{CHP}} \\
&= \begin{pmatrix} \frac{420 \cdot c_{\text{electricity}}}{1000} \\ 0 \\ 0 \\ 0 \end{pmatrix},
\end{aligned} \tag{A.13}
$$

$$
\begin{aligned}
\mathbf{g}_{\text{CHP, no loop}} &= \mathbf{B}_{\text{CHP, no loop}} \cdot \mathbf{A}^{-1}_{\text{CHP, no loop}} \cdot \mathbf{f}_{\text{CHP}} \\
&= \begin{pmatrix} \frac{-504 \cdot c_{\text{electricity}}}{1847} \end{pmatrix},
\end{aligned} \tag{A.14}
$$

$$
\begin{aligned}
\mathbf{h}_{\text{CHP, no loop}} &= \mathbf{Q}_{\text{CHP}} \cdot \mathbf{B}_{\text{CHP, no loop}} \cdot \mathbf{A}^{-1}_{\text{CHP, no loop}} \cdot \mathbf{f}_{\text{CHP}} \\
&= \begin{pmatrix} \frac{504 \cdot c_{\text{electricity}}}{1847} \end{pmatrix}.
\end{aligned} \tag{A.15}
$$

Appendix B

Student theses completed within this work

The following student theses were completed within the scope of this work:

Assen, N. v.d. (2010). Vergleich alternativer Methoden zur Berücksichtigung von Unsicherheiten in Ökobilanzen (in German). Diploma thesis, RWTH Aachen University.

Eichstädt, P. (2012). Berechnung von Ökobilanzen multifunktionaler Prozesssysteme in Matlab (in German). Seminar thesis, RWTH Aachen University.

Leder, J. (2009). Ökobilanzieller Vergleich zweier Membran-Verfahren zur Chlorherstellung (in German). Seminar thesis, RWTH Aachen University.

Bibliography

Aldrich, R., F. X. Llauró, J. Puig, P. Mutjé, and M. Àngels Pàlach (2011). Allocation of GHG emissions in combined heat and power systems: a new proposal for considering inefficiencies of the system. *J. Cleaner Prod. 19*(9), 1072 – 1079.

American Chemistry Council (2013). The chlorine tree. Chlorine Chemistry Division of the American Chemistry Council. Available online: http://www.chlorinetree.org, last accessed January 18th, 2013.

Arbeitsgemeinschaft Energiebilanzen (2013). Bruttostromerzeugung in Deutschland von 1990 bis 2012 nach Energieträgern (in German). Technical report. Arbeitsgemeinschaft Energiebilanzen e.V.. Available online: http://www.ag-energiebilanzen.de (last accessed: April 15th 2013).

Ardente, F. and M. Cellura (2012). Economic allocation in life cycle assessment. *J. Ind. Ecol. 16*(3), 387–398.

Ayer, N. W., P. H. Tyedmers, N. L. Pelletier, U. Sonesson, and A. Scholz (2007). Co-product allocation in life cycle assessments of seafood production systems: Review of problems and strategies. *Int. J. LCA 12*(7), 480–487.

Ayres, R. U. (1997). Metals recycling: economic and environmental implications. *Resour. Conserv. Recycl. 21*(3), 145 – 173.

Azapagic, A. (1996). *Environmental System Analysis: The application of linear programming to life cycle assessment.* Ph. D. thesis, University of Surrey.

Azapagic, A. and R. Clift (1999a). Allocation of environmental burdens in co-product systems: Product-related burdens (Part 1). *Int. J. LCA 4*, 357–369.

Azapagic, A. and R. Clift (1999b). Allocation of environmental burdens in multiple-function systems. *J. Cleaner Prod. 7*(2), 101 – 119.

Azapagic, A. and R. Clift (2000). Allocation of environmental burdens in co-product systems: Process and product-related burdens (part 2). *Int. J. LCA 5*, 31–36.

Babusiaux, D. and A. Pierru (2007). Modelling and allocation of CO2 emissions in a multiproduct industry: The case of oil refining. *Appl. Energy 84* (7-8), 828–841.

Baker, J. and M. Lepech (2009). Treatment of uncertainties in life cycle assessment. In *Proceedings of the 10th International Conference on Structural Safety and Reliability (ICOSSAR09), Osaka, Japan.*

Baumann, H. and A.-M. Tillman (2004). *The Hitch Hiker's Guide to LCA. An orientation in LCA Methodology and Application.* Studentlitteratur AB, Lund, Sweden.

Bayer (2011). Sustainable development report 2011. Technical report, Bayer AG, Leverkusen. Available online: http://www.sustainability2011.bayer.com/, last accessed May 16th 2013.

Bayer Material Science (2010a). Personal communication. March 29th 2010.

Bayer Material Science (2010b). Personal communication. September 7th 2010.

Bayer Material Science (2012). Personal communication. March 9th 2012.

Ben-Israel, A. and T. Greville (2003). *Generalized Inverses: Theory and Applications.* CMS Books in Mathematics. Springer, New York, United States.

Benetto, E., C. Dujet, and P. Rousseaux (2006). Possibility theory: A new approach to uncertainty analysis? *Int. J. LCA 11*, 114–116.

Bernesson, S., D. Nilsson, and P. Hansson (2004). A limited LCA comparing large- and small-scale production of rape methyl ester (RME) under swedish conditions. *Biomass Bioenergy 26* (6), 545–559.

Bernstein, L., J. Roy, K. C. Delhotal, J. Harnisch, R. Matsuhashi, L. Price, K. Tanaka, E. Worrell, F. Yamba, and Z. Fengqi (2007). *Climate Change 2007: Mitigation. Contribution of Working Group III to the Fourth Assessment Report of the Intergovernmental Panel on Climate Change*, Chapter Industry, pp. 447–496. Cambridge University Press, Cambridge, United Kingdom.

Bier, J., C. Verbeek, and M. Lay (2012). An eco-profile of thermoplastic protein derived from blood meal Part 1: allocation issues. *Int. J. LCA 17*, 208–219.

Björklund, A. (2002). Survey of approaches to improve reliability in LCA. *Int. J. LCA 7*, 64–72.

Bojarski, A. D., G. Guillen-Gosalbez, L. Jimenez, A. Espuna, and L. Puigjaner (2008). Life cycle assessment coupled with process simulation under uncertainty for reduced environmental impact: Application to phosphoric acid production. *Ind. Eng. Chem. Res. 47*(21), 8286–8300.

Boustead, I. (1972). *The milk bottle.* Open University Press, Milton Keynes, United Kingdom.

Boustead, I. (1996). LCA - How it came about - The beginning in the U.K. *Int. J. LCA 1*, 147–150.

British Standards Institution (2011). PAS 2050:2011 Specification for the assessment of the life cycle greenhouse gas emissions of goods and services. Britisch Standards Institution. Available online: http://shop.bsigroup.com/Navigate-by/PAS/, last accessed May 16th 2013.

Börjeson, L., M. Höjer, K.-H. Dreborg, T. Ekvall, and G. Finnveden (2006). Scenario types and techniques: Towards a user's guide. *Futures 38*(7), 723 – 739.

Bulan, A., J. Jörissen, J. Jung, R. Kiefer, H. Lochhaas, J. Röttger-Heinz, T. Turek, and N. Wagner (2009). *Klimaschutz und Anpassung an die Klimafolgen - Strategien, Maßnahmen, und Anwendungsbeispiele*, Chapter CO2-Reduktion bei der Herstellung chemischer Grundstoffe - Einsatz von Sauerstoffverzehrkathoden bei der Chlorherstellung (in German), pp. 61–67. Institut der deutschen Wirtschaft, Köln, Germany.

Bundesverband Energie- und Wasserwirtschaft (2013). Erdgasbezugsquellen: Deutsches Erdgasaufkommen nach Herkunftsländern (in German). German Association of Energy and Water Industries. Available online: http://bdew.de/internet.nsf/id/daten-grafik-de, last accessed May 16th 2013.

Cattlin, J. R. and Y. Wang (2012). Recycling gone bad: when the option to recycling increases resources consumption. *J. Consum. Psychol. 23*, 122–127.

Cederberg, C. and M. Stadig (2003). System expansion and allocation in life cycle assessment of milk and beef production. *Int. J. LCA 8*(6), 350–356.

Chevalier, J.-L. and J.-F. Tano (1996). Life cycle analysis with ill-defined data and its application to building products. *Int. J. LCA 1*, 90–96.

Ciroth, A. (2001). *Fehlerrechnung in Ökobilanzen (in German)*. Ph. D. thesis, Prozesswissenschaften der Technischen Universität Berlin.

Ciroth, A., G. Fleischer, and J. Steinbach (2004). Uncertainty calculation in life cycle assessments. *Int. J. LCA 9*, 216–226.

Cooper, J., C. Godwin, and E. Hall (2008). Modeling process and material alternatives in life cycle assessments. *Int. J. LCA 13*, 115–123.

Cruze, N., P. Goel, and B. Bakshi (2013). On the "rigorous proof of fuzzy error propagation with matrix-based LCI". *Int. J. LCA*, 1–4.

Curran, M. A. (2007a). Co-product and input allocation approaches for creating life cycle inventory data: A literature review. *Int. J. LCA 12*, 65–78.

Curran, M. A. (2007b). Studying the effect on system preference by varying coproduct allocation in creating life-cycle inventory. *Environ. Sci. Technol. 41* (20), 7145–7151.

De La Mantia, F. and F. Mantia (2002). *Handbook of plastics recycling*. iSmithers Rapra Publishing, Shawbury, United Kinddom.

Ekvall, T. (1999a). Comment on the marginal approach to allocation. *J. Cleaner Prod. 7*, 465–466.

Ekvall, T. (1999b). *System Expansion and Allocation in Life Cycle Assessment*. Ph. D. thesis, Chalmers University of Technology.

Ekvall, T. and G. Finnveden (2001). Allocation in ISO 14041–a critical review. *J. Cleaner Prod. 9* (3), 197 – 208.

El-Halwagi, M. (2006). *Process Integration*. Process Systems Engineering. Academic Press, Waltham, United States.

EuroChlor (2011). Key facts about chlorine. Technical report, Euro Chlor. Available online: http://www.eurochlor.org/download-centre/key-fact-cards.aspx, last accessed May16th 2013.

EuroChlor (2012). Chlorine Industry Review 2011-2012. Technical report, Euro Chlor. Available online: http://www.eurochlor.org/download-centre/the-chlorine-industry-review.aspx, last accessed May16th 2013.

European Commission (2001). Reference document on best-available techniques in chlor-alkali manufacturing industry. Technical report, European Commission - Integrated Pollution Prevention and Control (IPPC) - Institute for Prospective Technological Studies. Available online: http://eippcb.jrc.ec.europa.eu//reference/, last accessed May 16th 2013.

European Commission (2010a). Analysis of existing environmental impact assessment methodologies for use in Life Cycle Assessment. ILCD Handbook - International Reference Life Cycle Data System, European Comission - Joint Research Centre - Institute for Environment and Sustainability. Available online: http://lct.jrc.ec.europa.eu/assessment/publications, last accessed May 16th 2013.

European Commission (2010b). International Reference Life Cycle Data System (ILCD) Handbook - General guide for Life Cycle Assessment - Detailed Guidance. Technical Report First edition, European Commission - Joint Research Centre - Institute for Environment and Sustainability - EUR 24708 EN. Luxembourg. Publications Office of the European Union. Available online: http://lct.jrc.ec.europa.eu/assessment/publications, last accessed May 16th 2013.

Fauvarque, J. (1996). The chlorine industry. *Pure Appl. Chem. 68*(9), 1713–1720.

Feitz, A. J., S. Lundie, G. Dennien, M. Morain, and M. Jones (2007). Generation of an industry-specific physico-chemical allocation matrix - Application in the dairy industry and implications for systems analysis. *Int. J. LCA 12*(2), 109–117.

Fink, P. (1997). The roots of LCA in switzerland - continuous learning by doing. *Int. J. LCA 2*, 131–134.

Finnveden, G. and T. Ekvall (1998). Life-cycle assessment as a decision-support tool– the case of recycling versus incineration of paper. *Resour. Conserv. Recycl. 24* (3-4), 235 – 256.

Finnveden, G., M. Z. Hauschild, T. Ekvall, J. Guinee, R. Heijungs, S. Hellweg, A. Koehler, D. Pennington, and S. Suh (2009). Recent developments in Life Cycle Assessment. *J. Environ. Manage. 91*(1), 1–21.

Flysö, A., C. Cederberg, M. Henriksson, and S. Ledgard (2011). How does co-product handling affect the carbon footprint of milk? Case study of milk production in new zealand and sweden. *Int. J. LCA 16*, 420–430.

Foo, D. C. (2012). *Process integration for resources conservation.* CRC Press, Boca Raton, United States.

Friedler, F. (2010). Process integration, modelling and optimisation for energy saving and pollution reduction. *Appl. Therm. Eng. 30*(16), 2270 – 2280.

Frischknecht, R. (1998). *Life cycle inventory analysis for decision-making. Scope-dependent inventory system models and context-specific joint product allocation.* Ph. D. thesis, Swiss Federal Institute of Technology Zürich.

Frischknecht, R. (2000). Allocation in life cycle inventory analysis for joint production. *Int. J. LCA 5*, 85–95.

Frischknecht, R., N. Jungbluth, H.-J. Althaus, G. Doka, R. Dones, T. Heck, S. Hellweg, R. Hischier, T. Nemecek, G. Rebitzer, and M. Spielmann (2005). The ecoinvent database: Overview and methodological framework. *Int. J. LCA 10*, 3–9.

Geisler, G., S. Hellweg, and K. Hungerbühler (2004). Uncertainties in lca of plant-growth regulators and implications on decision-making. In C. Pahl-Wostl, S. Schmidt, and T. Jakeman (Eds.), *iEMSs 2004 International Congress: "Complexity and Integrated Resources Management". International Environmental Modelling and Software Societey, Osnabrueck, Germany.*

Geisler, G., S. Hellweg, and K. Hungerbühler (2005). Uncertainty analysis in life cycle assessment (LCA): Case study on plant-protection products and implications for decision making. *Int. J. LCA 10*, 184–192.

Güereca, L., N. Agell, S. Gassa, and J. Baldasano (2007). Fuzzy approach to life cycle impact assessment. *Int. J. LCA 12*, 488–496.

Goedkoop, M., R. Heijungs, M. A. J. Huijbregts, A. de Schryver, J. Struijs, and R. van Zelm (2013). ReCiPe 2008 - A life cycle impact assessment method which comprises harmonised category indicators at the midpoint and the endpoint level. Technical report, Ministerie van Volkshuisvesting, Ruimtelijke Ordening en Milieubeheer. Available online: http://www.lcia-recipe.net, last accessed May 16th 2013.

Goedkoop, M., A. D. Schryver, and M. Oele (2006). Introduction to LCA with Simapro 7. Technical report, PRé Consultants. Available online: http://www.pre-sustainability.com/manuals. Last accessed May 10th 2013.

Gonzalez, A., J. Sala, I. Flores, and L. Lopez (2003). Application of thermoeconomics to the allocation of environmental loads in the life cycle assessment of cogeneration plants. *Energy 28*(6), 557–574.

González-García, S., M. T. Moreira, and G. Feijoo (2010). Comparative environmental performance of lignocellulosic ethanol from different feedstocks. *Renewable Sustainable Energy Rev. 14*(7), 2077 – 2085.

Graus, W. and E. Worrell (2011). Methods for calculating CO2 intensity of power generation and consumption: A global perspective. *Energy Policy 39*(2), 613 – 627.

Guinée, J., M. Gorée, R. Heijungs, G. Huppes, R. Kleijn, A. d. Koening, L. v Oers, A. Wegener Sleeswijk, S. Suh, H. Udo de Haes, H. d. Bruijn, R. v. Duin, and M. Huijbregts (2002). *Handbook on Life Cycle Assessment: Operational Guide to the ISO Standards*. Eco-Efficiency in Industry and Science. Kluwer Academic Publishers, Dordrecht, The Netherlands.

Guinée, J., R. Heijungs, and E. van der Voet (2009). A greenhouse gas indicator for bioenergy: some theoretical issues with practical implications. *Int. J. LCA 14*, 328–339.

Guinee, J., R. Heijungs, and G. Huppes (2004). Economic allocation: Examples and derived decision tree. *Int. J. LCA 9*(1), 23–33.

Guinee, J. B. and R. Heiijungs (2007). Calculating the influence of alternative allocation scenarios in fossil fuel chains. *Int. J. LCA 12*(3), 173–180.

Guinee, J. B., R. Heijungs, G. Huppes, A. Zamagni, P. Masoni, R. Buonamici, T. Ekvall, and T. Rydberg (2011). Life cycle assessment: Past, present, and future. *Environ. Sci. Technol. 45*(1), 90–96.

Habert, G. (2012). A method for allocation according to the economic behaviour in the eu-ets for by-products used in cement industry. *Int. J. LCA*, 1–14.

Hamann, C. H., A. Hamnett, and W. Vielstich (2007). *Electrochemistry* (2 ed.). Wiley-VCH, Weinheim, Germany.

Hanssen, O. J. and O. A. Asbjornsen (1996). Statistical properties of emission data in life cycle assessments. *J. Cleaner Prod. 4*(3-4), 149 – 157.

Heijungs, R. (1994). A generic method for the identification of options for cleaner products. *Ecol. Econ. 10*(1), 69 – 81.

Heijungs, R. (1996). Identification of key issues for further investigation in improving the reliability of life-cycle assessments. *J. Cleaner Prod. 4*(3-4), 159–166.

Heijungs, R. (1997). *Economic drama and the environmental stage: Formal derivation of algorithmic tools for environmental analysis and decision-support from a unified epistemological principle*. Ph. D. thesis, University of Leiden.

Heijungs, R. (2010). Sensitivity coefficients for matrix-based LCA. *Int. J. LCA 15*(5), 511–520.

Heijungs, R. (2013a). CMLCA software. http://www.cmlca.eu/. Last accessed May 10th 2013.

Heijungs, R. (2013b). Personal communication. Januar 24th 2013.

Heijungs, R. and R. Frischknecht (1998). A special view on the nature of the allocation problem. *Int. J. LCA 3*(6), 321–332.

Heijungs, R. and R. Frischknecht (2005). Representing Statistical Distributions for Uncertain Parameters in LCA. Relationships between mathematical forms, their representation in EcoSpold, and their representation in CMLCA. *Int. J. LCA 10*, 248–254.

Heijungs, R. and J. B. Guinee (2007). Allocation and 'what-if' scenarios in life cycle assessment of waste management systems. *Waste Management 27*(8), 997–1005.

Heijungs, R. and M. A. J. Huijbregts (2004). A review of approaches to treat uncertainty in lca. In C. Pahl-Wostl, S. Schmidt, and T. Jakeman (Eds.), *iEMSs 2004 International Congress: "Complexity and Integrated Resources Management". International Environmental Modelling and Software Societey, Osnabrueck, Germany*.

Heijungs, R. and R. Kleijn (2001). Numerical approaches towards life cycle interpretation - Five examples. *Int. J. LCA 6*(3), 141–148.

Heijungs, R. and S. Suh (2002). *The Computational Structure of Life Cycle Assessment*. Eco-Efficiency in Industry and Science. Kluwer Academic Publishers, Dordrecht, The Netherlands.

Heijungs, R. and R. R. Tan (2010). Rigorous proof of fuzzy error propagation with matrix-based LCI. *Int. J. LCA 15*(9), 1014–1019.

Hischier, R., P. Wäger, and J. Gauglhofer (2005). Does WEEE recycling make sense from an environmental perspective? The environmental impacts of the Swiss take-back and recycling systems for waste electrical and electronic equipment (WEEE). *Environ. Impact Assess. Rev. 25*(5), 525 – 539.

Hojer, M., S. Ahlroth, K.-H. Dreborg, T. Ekvall, G. Finnveden, O. Hjelm, E. Hochschorner, M. Nilsson, and V. Palm (2008). Scenarios in selected tools for environmental systems analysis. *J. Cleaner Prod. 16*(18), 1958–1970.

Hong, J., S. Shaked, R. Rosenbaum, and O. Jolliet (2010). Analytical uncertainty propagation in life cycle inventory and impact assessment: application to an automobile front panel. *Int. J. LCA 15*(5), 499–510.

Huijbregts, M. A. J. (1998a). Application of uncertainty and variability in LCA, Part 1. *Int. J. LCA 3*, 273–280.

Huijbregts, M. A. J. (1998b). Application of uncertainty and variability in LCA, Part 2. *Int. J. LCA 3*, 343–351.

Huijbregts, M. A. J. (2001). *Uncertainty and variability in environmental life-cycle assessment.* Ph. D. thesis, University of Amsterdam.

Huijbregts, M. A. J., W. Gilijamse, A. M. J. Ragas, and L. Reijnders (2003). Evaluating uncertainty in environmental life-cycle assessment. a case study comparing two insulation options for a dutch one-family dwelling. *Environ. Sci. Technol. 37*(11), 2600–2608.

Hung, M.-L. and H.-W. Ma (2009). Quantifying system uncertainty of life cycle assessment based on monte carlo simulation. *Int. J. LCA 14*, 19–27.

Hunt, R., W. Franklin, and R. Hunt (1996). LCA - How it came about - personal reflections on the origin and the development of LCA in the USA. *Int. J. LCA 1*, 4–7.

Hunt, R. and R. Welch (1974). Resource and environmental profile analysis of plastics and non-plastics containers. Technical report, The Society of the Plastics Industry, New York.

ifu (2013). Umberto software. http://www.umberto.de. Last accessed May 16th 2013.

Imbeault-Tétreault, H., O. Jolliet, L. Deschênes, and R. K. Rosenbaum (2013). Analytical propagation of uncertainty in life cycle assessment using matrix formulation. *J. Ind. Ecol.*. Article published first online.

International Energy Agency (2012). Key world energy statistics. Available online: http://www.iea.org/publications/freepublications/, last accessed February 11th 2013.

IPCC (2007). Climate Change 2007: The Physical Science Basis. Contribution of Working Group I to the Fourth Assessment Report of the Intergovernmental Panel on Climate Change. Technical report, Solomon, S., D. Qin, M. Manning, Z. Chen, M. Marquis, K.B. Averyt, M. Tignor and H.L. Miller (eds.). Cambridge University Press, Cambridge, United Kingdom.

ISO (2006a). ISO 14040:2006 Environmental Management - Life Cycle Assessment - Principles and Framework. International Organisation of Standardization, Brussels, Belgium.

ISO (2006b). ISO 14044:2006 Environmental Management - Life Cycle Assessment - Requirements and Guidelines. International Organisation of Standardization, Brussels, Belgium.

Jörissen, J., T. Turek, and R. Weber (2011). Chlorherstellung mit Sauerstoffverzehrkathoden. Energieeinsparung bei der Elektrolyse (in German). *Chemie in unserer Zeit 45*(3), 172–183.

Jung, J., N. von der Assen, and A. Bardow (2012). Allocation vs. system expansion: A systematic comparative framework using matrix-based sensitivity coefficients. In *Proceedings of the 10th Ecobalance Conference, 20th to 23rd November, Yokohama, Japan.*

Jung, J., N. von der Assen, and A. Bardow (2013a). Comparative LCA of multi-product processes with non-common products: a systematic approach applied to chlorine electrolysis technologies. *Int. J. LCA 18*(4), 828–839.

Jung, J., N. von der Assen, and A. Bardow (2013b). Sensitivity-coefficient based uncertainty analysis for multi-functionality in lca. *Int. J. LCA*. submitted for publication.

Jungmeier, G., F. Werner, A. Jarnehammar, C. Hohenthal, and K. Richter (2002a). Allocation in LCA of wood-based products - Experiences of Cost Action E9 - Part II. Examples. *Int. J. LCA 7*(6), 369–375.

Jungmeier, G., F. Werner, A. Jarnehammar, C. Hohenthal, and K. Richter (2002b). Allocation in LCA of wood-based products - Experiences of Cost Action E9 Part I. Methodology. *Int. J. LCA 7*(5), 290–294.

Kaufman, A. S., P. J. Meier, J. C. Sinistore, and D. J. Reinemann (2010). Applying life-cycle assessment to low carbon fuel standards - how allocation choices influence carbon intensity for renewable transportation fuels. *Energy Policy 38*(9), 5229 – 5241.

Kim, S. and B. Dale (2002). Allocation procedure in ethanol production system from corn grain - I. System expansion. *Int. J. LCA 7*(4), 237–243.

Kim, S. and M. Overcash (2000). Allocation procedure in multi-output process: an illustration of iso 14041. *Int. J. LCA 5*, 221–228.

Klemes, J. J., P. S. Varbanov, and S. Pierucci (2010). Process integration for energy and water saving, increasing efficiency and reducing environmental impact. *Applied Thermal Engineering 30*(16), 2265 – 2269.

Klöpffer, W. and B. Grahl (2009). *Ökobilanz (LCA): Ein Leitfaden Für Ausbildung und Beruf (in German)*. Wiley-VCH, Weinheim, Germany.

Leimkühler, H.-J., M. Wolf, B. Himmelreich, J. Korte, N. Steeghs, M. Röwenstrunk, and S. Mütze-Niewöhner (2010). *Managing CO2-Emissions in the Chemical Industry*. Wiley-VCH, Weinheim, Germany.

Lloyd, S. and R. Ries (2007). Characterizing, propagating, and analyzing uncertainty in life-cycle assessment: A survey of quantitative approaches. *J. Ind. Ecol. 11*(1), 161–179.

Lucas, K. (2000). On the thermodynamics of cogeneration. *Int. J. Therm. Sci. 39*(9-11), 1039–1046.

Lund, H. (2001). *McGraw-Hill Recycling Handbook* (2 ed.). McGraw-Hill handbooks. McGraw-Hill Professional, New York, United States.

Luo, L., E. van der Voet, G. Huppes, and H. A. Udo de Haes (2009). Allocation issues in LCA methodology: a case study of corn stover-based fuel ethanol. *Int. J. LCA 14*, 529–539.

Magnus, J. R. and H. Neudecker (2007). *Matrix Differential Calculus with Applications in Statistics and Economics* (3 ed.). Wiley-VCH, Weinheim, Germany.

Malca, J. and F. Freire (2006). Renewability and life-cycle energy efficiency of bioethanol and bio-ethyl tertiary butyl ether (bioETBE): Assessing the implications of allocation. *Energy 31*(15, Sp. Iss. SI), 3362–3380.

Margni, M., T. Gloria, J. Bare, J. Seppälä, B. Steen, J. Struijs, and O. Toffoletto, L andJolliet (2008). Guidance on how to move from current practice to recommended practice in life cycle impact assessment. Technical report, UNEP/SETAC Life Cycle Initiative. Available online: http://lcinitiative.unep.fr/includes/file.asp?site=lcinit&file=E8C5CAD7-D1AC-49BE-BBFF-7AA6D25F1DA4, last accessed February 3rd, 2013.

Marvuglia, A., M. Cellura, and R. Heijungs (2010). Toward a solution of allocation in life cycle inventories: the use of least-squares techniques. *Int. J. LCA 15*, 1020–1040.

Mattila, T., P. Leskinen, S. Soimakallio, and S. Sironen (2012). Uncertainty in environmentally conscious decision making: beer or wine? *Int. J. LCA 17*, 696–705.

Mattila, T., K. Marjukka, H. Dahlbo, R. Soukka, and T. Myllymaa (2011). Uncertainty and sensitivity in the carbon footprint of shopping bags. *J. Ind. Ecol. 15*(2), 217–227.

Maurice, B., R. Frischknecht, V. Coelho-Schwirtz, and K. Hungerbühler (2000). Uncertainty analysis in life cycle inventory. application to the production of electricity with french coal power plants. *J. Cleaner Prod. 8*(2), 95 – 108.

Mauris, G., V. Lasserre, and L. Foulloy (2001). A fuzzy approach for the expression of uncertainty in measurement. *Measurement 29*(3), 165–177.

McCleese, D. and P. LaPuma (2002). Using monte carlo simulation in life cycle assessment for electric and internal combustion vehicles. *Int. J. LCA 7*, 230–236.

Moore, E. (1920). On the reciprocal of the general algebraic matrix. *Bulletin of the American Mathematical Society 26*(9), 394–395.

Morgan, M. and M. Henrion (1992). *Uncertainty: A Guide to Dealing with Uncertainty in Quantitative Risk and Policy Analysis*. Cambridge University Press, Cambridge, United Kingdom.

Moussallem, I., J. Jörissen, U. Kunz, S. Pinnow, and T. Turek (2008). Chlor-alkali electrolysis with oxygen depolarized cathodes: history, present status and future prospects. *J. Appl. Electrochem. 38*, 1177–1194.

Moussallem, I., S. Pinnow, and T. Turek (2009). Möglichkeiten zur Energierückgewinnung aus Wasserstoff bei der Chlor-Alkali-Elektrolyse (in German). *Chem. Ing. Tech. 81*(4), 489–493.

Neelis, M., M. Patel, K. Blok, W. Haije, and P. Bach (2007). Approximation of theoretical energy-saving potentials for the petrochemical industry using energy balances for 68 key processes. *Energy 32*(7), 1104 – 1123.

Nguyen, T. L. T. and J. E. Hermansen (2012). System expansion for handling co-products in lca of sugar cane bio-energy systems: Ghg consequences of using molasses for ethanol production. *Appl. Energy 89*(1), 254 – 261.

Oberbacher, B., H. Nikodem, and W. Klöpffer (1996). Lca - how it came about - an early systems analysis of packaging for liquids. *Int. J. LCA 1*, 62–65.

Oberbacher, B., W. Schonborn, H. Czabon, H. Deibig, H. Hampel, W. Klöpffer, W. Wehr, S. Gwifssner, D. Hums, K. Lippe, D. Merz, R. Rasch, H. Nikodem, L. Lichtwer, and W. Schyschka (1974). *Degradable Plastics and Waste Problems.* Erich Schmidt Verlag, Berlin, Germany.

O'Brien, T., T. Bommaraju, and F. Hine (2005a). *Handbook of Chlor-Alkali Technology*, Chapter Introduction, pp. 1–16. Springer, New York, United States.

O'Brien, T., T. Bommaraju, and F. Hine (2005b). *Handbook of Chlor-Alkali Technology*, Chapter Chlor-Alkali Technologies, pp. 387–442. Springer, New York, United States.

O'Brien, T., T. Bommaraju, and F. Hine (2005c). *Handbook of Chlor-Alkali Technology*, Chapter Chemistry and Electrochemistry of the Chlor-Alkali Process, pp. 75–386. Springer, New York, United States.

O'Brien, T., T. Bommaraju, and F. Hine (2005d). *Handbook of Chlor-Alkali Technology*, Chapter History of the Chlor-Alkali Industry, pp. 17–35. Springer, New York, United States.

O'Brien, T., T. Bommaraju, and F. Hine (2005e). *Handbook of Chlor-Alkali Technology*, Chapter Future Developments, pp. 1463–1490. Springer, New York, United States.

O'Brien, T., T. Bommaraju, and F. Hine (2005f). *Handbook of Chlor-Alkali Technology*, Chapter Product Handling, pp. 765–1012. Springer, New York, United States.

PE International (2013). GaBi software. http://www.gabi-software.com, last accessed May 16th 2013. Last accessed May 16th 2013.

Pelletier, N. and P. Tyedmers (2011). An ecological economic critique of the use of market information in life cycle assessment research. *J. Ind. Ecol. 15*(3), 342–354.

Pelletier, N. and P. Tyedmers (2012). Response to Weinzettel. *J. Ind. Ecol. 16*(3), 456–458.

Pennington, D., J. Potting, G. Finnveden, E. Lindeijer, O. Jolliet, T. Rydberg, and G. Rebitzer (2004). Life cycle assessment Part 2: Current impact assessment practice. *Environ. Int. 30*(5), 721–739.

Penrose, R. (1955). A generalized inverse for matrices. *Mathematical Proceedings of the Cambridge Philosophical Society 51*, 406–413.

Peters, G. (2007). Efficient algorithms for life cycle assessment, input-output analysis, and monte-carlo analysis. *Int. J. LCA 12*, 373–380.

Petersen, K. and M. Pedersen (2012). The Matrix Cookbook. Available online: http://http://www2.imm.dtu.dk/pubdb/public/index.php. Last accessed May 16th 2013.

Pinnow, S., N. Chavan, and T. Turek (2011). Thin-film flooded agglomerate model for silver-based oxygen depolarized cathodes. *J. Appl. Electrochem. 41*, 1053–1064.

Pohl, C., M. Ros, B. Waldeck, and F. Dinkel (1996). *Life Cycle Assessment (LCA) - Quo Vadis?*, Chapter Imprecision and Uncertainty in LCA, pp. 51–68. Birkhaeuser, Basel, Switzerland.

Pohl, C. E. (1999). *Auch zu präzis ist ungenau! Unsicherheitsanalyse in Ökobilanzen und Alternativen zu "use many methods" (in German)*. Ph. D. thesis, Swiss Federal Institute of Technology Zürich.

PRé Consultants (2013). Simapro software. http://www.pre-sustainability.com/simapro-lca-software. Last accessed May 16th 2013.

Quinteiro, P., A. Araújo, A. C. Dias, B. Oliveira, and L. Arroja (2012). Allocation of energy consumption and greenhouse gas emissions in the production of earthenware ceramic pieces. *J. Cleaner Prod. 31*(0), 14 – 21.

Reap, J., F. Roman, S. Duncan, and B. Bras (2008a). A survey of unresolved problems in life cycle assessment. *Int. J. LCA 13*(5), 374–388.

Reap, J., F. Roman, S. Duncan, and B. Bras (2008b). A survey of unresolved problems in life cycle assessment - Part 1: goal and scope and inventory analysis. *Int. J. LCA 13*(4), 290–300.

Rebitzer, G., T. Ekvall, R. Frischknecht, D. Hunkeler, G. Norris, T. Rydberg, W. Schmidt, S. Suh, B. Weidema, and D. Pennington (2004). Life cycle assessment Part 1: Framework, goal and scope definition, inventory analysis, and applications. *Environ. Int. 30*(5), 701–720.

Ros, M. (1998). *Unsicherheit und Fuzziness in ökologischen Bewertngen- Orientierungen zu einer robusten Praxis der Ökobilanzierung (in German)*. Ph. D. thesis, Swiss Federal Institute of Technology Zürich.

Rosen, M. A. (2008). Allocating carbon dioxide emissions from cogeneration systems: descriptions of selected output-based methods. *J. Cleaner Prod. 16*(2), 171–177.

Sakai, S. and K. Yokoyama (2002). Formulation of sensitivity analysis in life cycle assessment using a perturbation method. *Clean Technol. Environ. Policy 4*, 72–78.

Sattler, K. (2001). *Thermische Trennverfahren*, Chapter Lösungsmitteleindampfung (in German), pp. 617–676. Wiley-VCH, Weinheim, Germany.

Sayagh, S., A. Ventura, T. Hoang, D. François, and A. Jullien (2010). Sensitivity of the LCA allocation procedure for BFS recycled into pavement structures. *Resour. Conserv. Recycl. 54*(6), 348 – 358.

Schmidt, M. and A. Schorb (1996). *Stoffstromanalysen in Ökobilanzen und Öko-Audits (in German)*. Springer, Berlin, Germany.

Schmittinger, P., L. Calvert Curlin, T. Asawa, S. Kotowski, H. Beer, A. Greenberg, E. Zelfel, and B. R.W (2006). *Ullmann's Encyclopedia of Industrial Chemistry* (6th ed.)., Chapter Chlorine, pp. 399–481. Wiley-VCH, Weinheim, Germany.

Seppälä, J. (2007). On the meaning of fuzzy approach and normalisation in life cycle impact assessment. *Int. J. LCA 12*, 464–469.

Sills, D. L., V. Paramita, M. J. Franke, M. C. Johnson, T. M. Akabas, C. H. Greene, and J. W. Tester (2013). Quantitative uncertainty analysis of life cycle assessment for algal biofuel production. *Environ. Sci. Technol. 47*(2), 687–694.

Soimakallio, S. and L. Saikku (2012). CO2 emissions attributed to annual average electricity consumption in oecd (the organisation for economic co-operation and development) countries. *Energy 38*(1), 13 – 20.

Sonnemann, G., M. Schuhmacher, and F. Castells (2003). Uncertainty assessment by a Monte Carlo simulation in a life cycle inventory of electricity produced by a waste incinerator. *J. Cleaner Prod. 11*(3), 279–292.

Sonnemann, G. and B. Vigon (2011). Global guidance principles for life cycle assessment databases. UNEP/SETAC Life Cycle Initiative. Available online: http://lcinitiative.unep.fr/includes/file.asp?site=lcinit&file=E8C5CAD7-D1AC-49BE-BBFF-7AA6D25F1DA4 (last accessed February 3rd, 2013).

Spielmann, M., R. Scholz, O. Tietje, and P. d. Haan (2005). Scenario modelling in prospective lca of transport systems. application of formative scenario analysis. *Int. J. LCA 10*, 325–335.

Sugiyama, M., K. Saiki, A. Sakata, H. Aikawa, and N. Furuya (2003). Accelerated degradation testing of gas diffusion electrodes for the chlor-alkali process. *J. Appl. Electrochem. 33*, 929–932.

Suh, S. and G. Huppes (2005). Methods for life cycle inventory of a product. *J. Cleaner Prod. 13*(7), 687 – 697.

Svanes, E., M. Vold, and O. Hanssen (2011). Effect of different allocation methods on LCA results of products from wild-caught fish and on the use of such results. *Int. J. LCA 16*, 512–521.

Tan, R. (2008). Using fuzzy numbers to propagate uncertainty in matrix-based LCI. *Int. J. LCA 13*, 585–592.

Taylor, J. (1982). *An introduction to error analysis. The study of uncertainties in physical measurements.* University Science Books, Mill Valley, United States.

Thrane, M. (2004). Energy consumption in the danish fishery: Identification of key factors. *J. Ind. Ecol. 8*(1-2), 223–239.

Tillman, A.-M., T. Ekvall, H. Baumann, and T. Rydberg (1994). Choice of system boundaries in life cycle assessment. *J. Cleaner Prod. 2*(1), 21–29.

Udo de Haes, H. A. and M. van Rooijen (2005). Life cycle approaches - the road from analysis to practice. Technical report, UNEP-SETAC Life Cycle Initiative. Available online: http://www.unep.fr/scp/publications/details.asp?id=DTI/0594/PA, last accessed May 16th 2013.

Ursua, A., L. Gandia, and P. Sanchis (2012). Hydrogen production from water electrolysis: Current status and future trends. *Proceedings of the IEEE 100*(2), 410 –426.

von der Assen, N. (2011). Schematic Figure of a Product Life Cycle. RWTH Aachen University, unpublished.

Vose, D. (1996). *Quantitative Risk Analysis: A Guide to Monte Carlo Simulation Modelling*. John Wiley & Sons, Chichester, New York, Brisbane, Toronto, Singapore.

Wardenaar, T., T. van Ruijven, A. Beltran, K. Vad, J. Guinée, and R. Heijungs (2013). Differences between LCA for analysis and LCA for policy: a case study on the consequences of allocation choices in bio-energy policies. *Int. J. LCA*, 1–9.

Weidema, B. (2000). Avoiding co-product allocation in life-cycle assessment. *J. Ind. Ecol. 4*(3), 11–33.

Weidema, B. P. and J. H. Schmidt (2010). Avoiding allocation in life cycle assessment revisited. *J. Ind. Ecol. 14*(2), 192–195.

Weinzettel, J. (2012). Understanding who is responsible for pollution: What only the market can tell us - comment on "an ecological economic critique of the use of market information in life cycle assessment research". *J. Ind. Ecol. 16*(3), 455–456.

Werner, F. and K. Richter (2000). Economic allocation in LCA: A case study about aluminium window frames. *Int. J. LCA 5*, 79–83.

World Resources Institute (2010). Greenhouse Gas Protocol - Product Life Cycle Accounting and Reporting Standard. Technical report, World Resources Institute - World Business Council for Sustainable Development.

Yeager, E. and P. Bindra (1980). Sauerstoff-Verzehrkathoden für die Chloralkali-Elektrolyse (in German). *Chem. Ing. Tech. 52*(5), 384–391.

Aachener Beiträge zur Technischen Thermodynamik

ABTT 1
Phillip Voll
Automated Optimization-Based Synthesis of Distributed Energy Supply Systems
1. Auflage 2014
ISBN 978-3-86130-474-6

ABTT 2
Johannes Jung
Comparative Life Cycle Assessment of Industrial Multi-Product Processes
1. Auflage 2014
ISBN 978-3-86130-471-5